simply shetland 4

at Tomales Bay

The Smith Brothers Fishery

Tomales and Bodega Bays both have long fishing histories—for centuries by the Coast Miwok—more recently by Mexicans and Americans. In time, the Mexicans would cede the land to the Americans, who would force the Miwok northward. The Smith family would encompass all three and become the predominant fishing family in the area.

In 1841, Captain Stephen Smith of Boston sailed north from San Francisco Bay, married Peruvian Manuela Torres, became a Mexican citizen and was thus able to obtain, through a land grant, the 35,000-acre Rancho Bodega—bordered by the Russian River and Estero Americano. This made him a very powerful man, indeed, as it gave him control over a navigable body of water with an outlet to the Pacific Ocean. Over the next few years, he built a warehouse at Bodega Bay and established the first regular ocean freight and passenger service between Bodega and San Francisco. This, in turn, brought ranchers to the area, who could more easily ship their goods.

Captain Smith's relationship with a Coast Miwok woman named Tsupu would link the Smith family with the Miwok people. In 1844, a son was born, William (his Miwok name was Yole Tamal), who, in time, also married a Miwok woman. When the militia forced the Miwok out of the area in the 1860's, William accompanied his wife on the march up to Mendocino, where she would die. Heartbroken, but with a strong sense of purpose, William returned to Bodega Bay. He worked hard and fished out of a small sailboat he built himself. Later, he married Rosalie Charles and the two had 11 children. The family operated two boats out of Bodega Bay—the *Six Brothers* and the *Five Sisters*, named for the Smith children. In 1911, five of the brothers joined a commercial fishing company for the king crab season. This proved to be extremely profitable. Upon their return, the story goes, "they filled their mother's apron with $20 gold pieces." This was enough money to build two 50-foot boats, christened *Smith Brothers No. 1* and *No. 2*—state of the art for the time. By the 1930's the Smith Brothers Fishery had the largest fleet in Bodega Bay and fished in all seasons for salmon, bottom fish, crab, and herring. The sisters helped with the processing, filleting flounder and sole. This long-lived family business ended in the 1970's.

To learn more about Tomales and Bodega Bay history, visit WWW.TOMALESHISTORY.COM.

Crew of Smith Brothers No. 1 (left to right) Eddy Smith, Bill Orr, Harold Ames & Bill Smith. Our thanks to the Tomales Regional History Center.

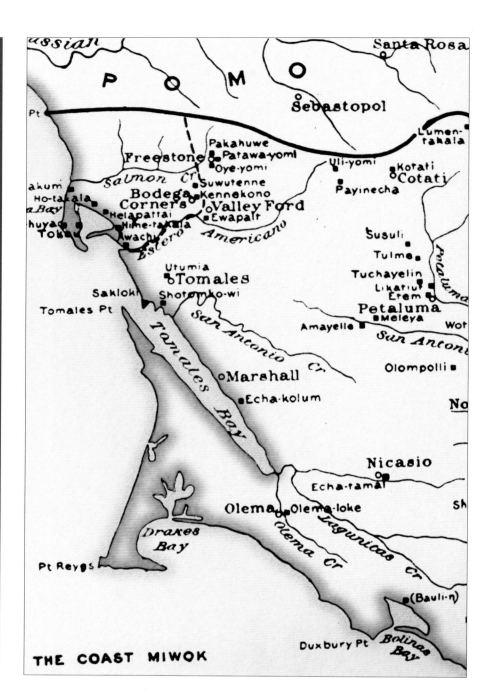

THE COAST MIWOK

introduction

Heading north from San Francisco across the Golden Gate Bridge, most people stay inland on the fast road. Better to take your time and turn to the west towards the Pacific Ocean on old Highway 1. The scenery can't be matched as you wind and twist up the Shoreline Highway, past Point Reyes and eventually reach the crossroads that is Tomales. Once a fishing town, today it is a busy stopover and supply post for cyclists, sightseers and beachgoers.

Tomales Bay's casual charm makes it the perfect setting for our fall knitting book. As always, we offer traditional Shetland wools knitted up into clothes with a modern, casual look. Eunny Jang's stunning *Autumn Rose Pullover* takes Fair Isle knitting into the new century, with fitted shaping and a compound angle raglan sleeve. Carol Lapin's tiered *Drake's Bay Jacket & Coat* and *Tomales Bay Skirt* once again use traditional yarns in modern ways.

This year, we introduce two new yarns. Blended with luxury fibers, both retain the rusticity, coloring and life of traditional Shetland wool while providing a beautifully rich and unexpected "hand." Our SILK & SHETLAND LAMBSWOOL version of the *Duxbury Point Pullover* turns a finely knit aran garment into a work of elegance. The character of our SHETLAND LAMBSWOOL & CASHMERE is beautifully showcased in the *Olema Turtleneck Pullover and Hat*. It is a knitting yarn with all the character expected of Scottish wool with enough cashmere to allow the cozy turtleneck design.

This year's collection is by far the most varied of our books. Though we will always love the technique and beauty of an Aran shawl, we also enjoy the excitement of a design like *Passive Polka Dots*. Enjoy your trip around Tomales Bay. We're sure you'll find something fun to knit and wonderful to wear for years to come.

DIEKMANN'S GENERAL STORE

Over the past century, a number of general stores have operated in the town of Tomales under various names and family ownerships. Diekmann's General Store, still in business today, remains as a reminder of the general store's vital role in small town life. Originally called Newburgh & Kahn's, it supplied local residents with salt, sugar, flour, coffee, hay and grain, piece goods, coal, gun powder and lumber. Sold in 1898 to Leon Dickinson, the store would remain in the Dickinson family for two generations. Leon's son, Bray, took over the store in 1936 and acted as both shopkeeper and postmaster.

Walt Diekmann eventually purchased the store and began another two-generation proprietorship. Acting on the encouragement of brother William, one of three brothers who all owned various stores in nearby towns, Walt and Mildred Diekmann packed their household goods in a one-wheeled trailer and set off from Iowa to California. All three Diekmann children—Bill, Mark and Kristen—would eventually work in the store, stocking shelves when they were young, and learning the finer points of storekeeping as they grew older. Walt Diekmann was beloved by local children, as he always had treats for them at Christmas-time and allowed liberal reading of comic books in the store.

Bill and Kristen took over the family business in 1972 and restored some of the original interior details and cabinetry. Today, under new ownership, the store remains a beloved part of Tomales life and is the "must stop" supply post for travelers along the coast highway.

PHOTOS—DIEKMANN GENERAL STORE TODAY (TOP); WALT & MILDRED DIEKMANN (BOTTOM).

Contents

introduction 3

Tomales History
the smith brothers fishery 2
diekmann's general store 4

Knitting Projects
ella jacket 6
olema turtleneck pullover 11
olema hat 14
duxbury point pullover 16
duxbury point wrap 22
passive polka dots 25
inverness cape 29
tomales bay skirt 33
tomales bay cowl 36
dillon beach pullover 41
alamere falls pullover 47
box stripe pullover 50
shadow pullover 59
drake's bay jacket & coat 62
rosalie scarf 68
leaf poncho jacket 70
carnivale cropped top 74
silk diamonds scarf 78
autumn rose pullover 82
glenice wrap 88

shade cards 92
abbreviations 95

ella jacket

 prudence makepeace

MATERIALS
YARN: Simply Shetland Silk & Lambswool (2 strands held together throughout) - 550 (600, 650) grams.
Shown in Craignish (030) on facing page and on pages 8 and 94.
NEEDLES: 40" circular US 8 (5 mm), *or correct needles to obtain gauge*.
ACCESSORIES: Six 5/8" buttons.

MEASUREMENTS
CHEST: 40 (43, 46)".
LENGTH: 20½ (21½, 22½)".
SLEEVE LENGTH: 16½ (17½, 17½)".

GAUGE
On US 8 in **Fisher Rib**: 16½ sts and 34 rows = 4".

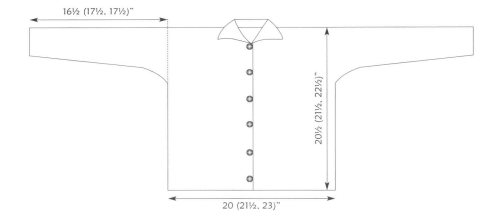

ella jacket

Designer Note
Make jacket in one piece beginning at right sleeve cuff. Work **Fisher Rib** for entire jacket, holding 2 strands of yarn together throughout. Where only one number is given, it applies to all sizes.

Fisher Rib (over odd number of sts)
Row 1 (RS) (Foundation Row): Edge st, *k1, p1; rep from * to last 2 sts; end k1, edge st.
Row 2 (WS): Edge st, *yo, sl 1 pwise, k1; rep from * to last 2 sts; end yo, sl 1 pwise, edge st.
Row 3 (RS): Edge st, *ktog the yo and sl st from previous row, p1; rep from * to last 3 sts (include yo as st); ktog the yo and sl st from previous row, edge st.

Rep Rows 2-3.

Jacket (Knitted Cuff-to-Cuff in One Piece)
Holding 2 strands of yarn together throughout, CO 41 sts. Work in **Fisher Rib** for 20 rows.

Shape Sleeve
Continue in **Fisher Rib** as set, inc'g 1 st at beg and end of next row 1 time, every 8th row 0 (0, 7) times, every 10th row 0 (8, 6) times, and every 12th row 9 (3, 0) times (61 (65, 69) sts on needle). CO 3 sts at end of next 12 rows (97 (101, 105) sts on needle).

Cast On for Body
CO 36 (38, 40) sts at end of next 2 rows (169 (177, 185) sts on needle). Work without further shaping until piece measures 6¾ (7½, 8¼)" from cast on for body, ending with RS facing for next row.

Divide Work
Next Row (RS): Work 85 (89, 93) sts and place on holder for right front; work rem 84 (88, 92) sts (for back) an additional 6¼", then place these sts onto holder.

Make Right Collar and Join to Right Front Stitches
With another US 8, CO 15 sts and work 21 rows in **Fisher Rib**, ending with WS facing for next row.

With WS's facing, place collar and right front sts onto same needle, making sure that collar sts are at collar end (top) of right front.

Next Row (WS): Continuing pattern as set, work collar sts, knitting tog last st of collar with first st of right front; work rem right front sts. Continue on these 99 (103, 107) sts for 3-1/8", ending with RS facing for next row.

Make Buttonholes
Next Row (RS): Work 3 (4, 3) sts, ([k2tog, yo, work 13 (13, 14) sts] 5 times); end k2tog, yo, work 19 (22, 22) sts.

Work 4 rows in pattern. BO.

Left Front
With US 8, CO 99 (103, 107) sts and work in **Fisher Rib** for 4", ending with RS facing for next row.

Next Row (RS): Work 85 (89, 93) sts; place last 14 sts on holder for left collar; continue on same needle as left front and knit sts from back holder (169 (177, 185) sts on needle).

Continue in pattern as set until piece measures 10½ (11½, 12¼) from cast on of left front. BO 36 (38, 40) sts at beg of next 2 rows (97 (101, 105) sts on needle). BO 3 sts at beg of next 12 rows (61 (65, 69) sts on needle). Then, reversing all shaping, work sleeve same as for right sleeve (dec at beg and end of every 12th row 9 (3, 0) times, every 10th row 0 (8, 6) times, and every 8th row 0 (0, 7) times), then dec 1 st at beg and end of next row. Work in pattern on rem 41 sts for 20 rows without further shaping. BO.

Work Left Collar
Place the 14 left collar sts from holder onto US 8, and continue collar in **Fisher Rib** as set, *inc'g 1 st at beg of 1st row to re-establish edge st.* When collar is same length as right front collar, BO.

Finishing
Join short edges of right and left collars and sew collar down along back neck edge. Sew side and sleeve seams. Sew on buttons opposite buttonholes. Block to finished measurements.

olema turtleneck pullover

betsy westman

MATERIALS
YARN: Simply Shetland Lambswool & Cashmere - 550 (600, 650, 700) grams. Shown in Kingfisher (453) on opposite page and on page 15.
NEEDLES: 24" circular US 6 (4 mm) and 16" circular US 5 (3.75 mm), *or correct needles to obtain gauge.*

MEASUREMENTS
CHEST: 40 (44, 48, 52)".
LENGTH TO UNDERARM: 14½ (14½, 14½, 14½)".
ARMHOLE DEPTH: 8 (8½, 9, 9½)".
LENGTH: 22½ (23, 23½, 24)".
SLEEVE LENGTH TO UNDERARM: 18 (18, 18½, 18½)".
SLEEVE LENGTH TO SHOULDER: 22 (22, 22½, 22½)".

GAUGE
On US 6 in st st: 23 sts and 32 rows = 4".

NOTE
Where only one number is given, it applies to all sizes.

NOTES ON CHART
Work odd-numbered (RS) rows from right to left, and even-numbered (WS) rows from left to right.

RIB PATTERN (MULTIPLE OF 3 + 2)
Row 1 (WS): P2, *k1, p2; rep from *.
Row 2 (RS): K2, *p1, k2; rep from *.

Rep Rows 1-2.

BACK
With US 6, CO 113 (125, 137, 149) sts. Work in **Rib Pattern** for 2", ending with RS facing for next row.

Next Row and All RS Rows: Knit.
All WS Rows: P2 (17, 5, 2), k1, ([p17, k1] 6 (5, 7 8) times); end p2 (17, 5, 2).

Continue as set until piece measures 14½" from CO edge, ending with RS facing for next row.

SHAPE UNDERARMS
BO 5 sts at beg of next 2 rows, then dec 1 st at beg and end of every row 4 times, then every other row 2 times. Continue on rem 91 (103, 115, 127) sts until armhole measures 7¾ (8¼, 8¾, 9¼)", ending with RS facing for next row.

SHAPE NECK AND SHOULDERS
Next Row (RS): Work 23 (29, 35, 41) sts; BO 45 sts for back neck; work rem 23 (29, 35, 41) sts.

Turn, and working each side separately, work 1 WS row, then BO.

FRONT
With US 6, CO 113 (125, 137, 149) sts. Work in **Rib Pattern** for 2", ending with RS facing for next row.

Next Row (RS): K46 (52, 58, 64) sts; place marker; k21, inc'g 5 sts evenly spaced across; place marker; knit rem 46 (52, 58, 64) sts (118 (130, 142, 154) sts on needle).

Next Row (WS): P2 (17, 5, 2); ([k1, p17] 2 (1, 2, 3) times), k1; p7 (16, 16, 7); work **Foundation Row** of **Cable Chart** over next 26 sts; p7 (16, 16, 7); k1, ([p17, k1] 2 (1, 2, 3) times); p2 (17, 5, 2).

olema turtleneck pullover

Next Row (RS): K46 (52, 58, 64); work Row 1 of **Cable Chart** over next 26 sts; k46 (52, 58, 64).

Continuing as set, rep the 32 rows of **Cable Chart** between markers and work background pattern as established until piece measures 14½" from CO edge, ending with RS facing for next row.

Shape Underarms
BO 5 sts at beg of next 2 rows, then dec 1 st at beg and end of every row 4 times, then every other row 2 times. Continue on rem 96 (108, 120, 132) sts until armhole measures 5 (5½, 6, 6½)", ending with RS facing for next row.

Shape Neck
Next Row (RS): Work 37 (43, 49, 55) sts; BO next 22 sts, dec'g 5 sts evenly spaced across; work rem 37 (43, 49, 55) sts.

Turn, and working each side separately, BO 4 sts at neck edge once, then dec at neck edge 1 st on every row 5 times, every 2nd row 4 times, and every 4th row 1 time. Continue in pattern as set on rem 23 (29, 35, 41) sts until same length as back. BO.

Sleeves
With US 6, CO 59 sts. Work in **Rib Pattern** for 2", ending with RS facing for next row.

Next Row and All RS Rows: Knit.
Next WS Row: P11, k1, p17, k1, p17, k1, p11.

Continue in pattern as set, **AND AT SAME TIME**, inc 1 st at beg and end of 3rd row 1 time, every 4th row 0 (0, 0, 6) times, every 6th row 0 (4, 14, 15) times, every 8th row 4 (11, 4, 0) times and every 10th row 8 (0, 0, 0) times, working inc'd sts into pattern (85 (91, 97, 103) sts on needle). Continue without further shaping until sleeve measures 18 (18, 18½, 18½)" ending with RS facing for next row.

Shape Sleeve Cap
BO 5 sts at beg of next 2 rows and 3 sts at beg of next 2 rows, then dec 1 st at beg and end of every row 5 times, every 2nd row 6 times, and every row 6 times (35 (41, 47, 53) sts on needle). BO 4 (5, 6, 7) sts at beg of next 4 rows. BO rem 19 (21, 23, 25) sts.

Join Shoulders
Sew shoulder seams.

Neckband
The neckband shown is a turtleneck. You'll be working in the rnd on the WS; when the neck is turned down, the RS will show. If you prefer a crewneck, work every rnd as: *k2, p1; rep from *.

With 16" circular US 5, RS facing, pick up 99 sts evenly around neck edge.

Every Rnd: *P2, k1; rep from *.

When neckband measures 6½", BO loosely in pattern.

Finishing
Sew sleeves to armholes. Sew side and sleeve seams. Weave in ends. Block to finished measurements.

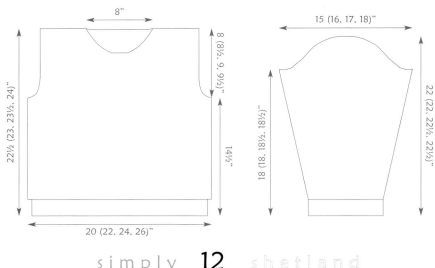

olema turtleneck pullover

Cable Chart

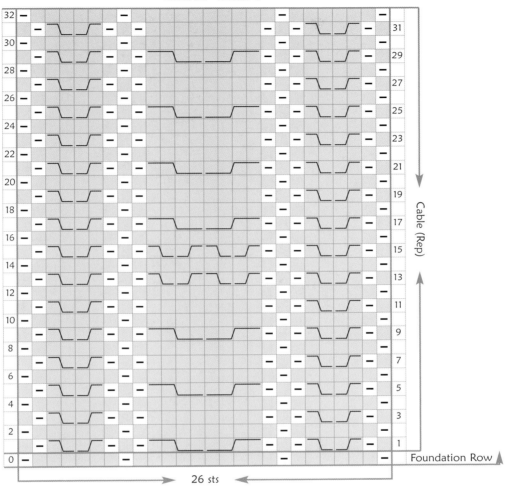

Key

- k on right side rows; p on wrong side rows.
- — p on right side rows; k on wrong side rows.
- sl 1 st to cn and hold at back; k1; k1 from cn.
- sl 1 st to cn and hold at front; k1; k1 from cn.
- sl 2 sts to cn and hold at back; k2; k2 from cn.
- sl 2 sts to cn and hold at front; k2; k2 from cn.

olema hat

betsy westman

MATERIALS
YARN: Simply Shetland Lambswool & Cashmere - 100 grams. Shown in Kingfisher (453).
NEEDLES: 16" circular and set of 4 double-pointed US 6 (4 mm), *or correct needles to obtain gauge.*

MEASUREMENTS
CIRCUMFERENCE: 21½". **LENGTH:** 8".

GAUGE
On US 6 in st st: 23 sts and 32 rows = 4".

NOTES ON CHARTS
Work odd-numbered (RS) rows from right to left, and even-numbered (WS) rows from left to right.

MAIN BODY
CO on 28 sts. Work **Chart** until piece measures 21½" from CO edge. BO. Sew cast-on and bound-off edges tog to form back seam.

BRIM
With RS facing, beg at at back seam, pick up 126 sts evenly spaced around bottom edge of main body, place marker, join and work in the rnd as follows:

Next 8 Rnds: K2, p1; rep from *.

BO in pattern.

CROWN
With circular needle, beg at back seam, pick up 126 sts evenly around top edge of main body, place marker, join and knit every rnd until hat measures 5¼" from beg of rib (only a few rows if any). Work dec rnds as follows, changing to double-pointed needles when sts get stretched.

Rnd 1 (Dec Rnd 1): ([K7, k2tog] 14 times) (112 sts rem).
Rnds 2, 3, 4, 6, 7, 8, 10, 11, 13, 14, 16, 18 & 20: Knit.
Rnd 5 (Dec Rnd 2): ([K6, k2tog] 14 times) (98 sts rem).
Rnd 9 (Dec Rnd 3): ([K5, k2tog] 14 times) (84 sts rem).
Rnd 12 (Dec Rnd 4): ([K4, k2tog] 14 times) (70 sts rem).
Rnd 15 (Dec Rnd 5): ([K3, k2tog] 14 times) (56 sts rem).
Rnd 17 (Dec Rnd 6): ([K2, k2tog] 14 times) (42 sts rem).
Rnd 19 (Dec Rnd 7): ([K1, k2tog] 14 times) (28 sts rem).
Rnd 21 (Dec Rnd 8): ([K2tog]) 14 times) (14 sts rem).
Rnd 22 (Dec Rnd 9): ([K2tog] 7 times) (7 sts rem).
Rnd 23 (Dec Rnd 10): ([K2tog] twice), k3tog (3 sts rem).

Break yarn, thread onto darning needle, draw through rem sts, pull through hole to inside and fasten off. Weave in ends.

Key
- k on right side rows; p on wrong side rows.
- p on right side rows; k on wrong side rows.
- sl 1 st to cn and hold at back; k1; k1 from cn.
- sl 1 st to cn and hold at front; k1; k1 from cn.
- sl 2 sts to cn and hold at back; k2; k2 from cn.
- sl 2 sts to cn and hold at front; k2; k2 from cn.

Chart

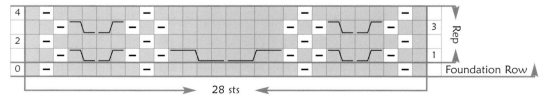

duxbury point pullover

beatrice smith

Materials
Yarn: In Simply Shetland Silk & Lambswool - 550 (600, 700) grams; shown in Glenbuchat (033) on pages 17 & 20. In Jamieson's Shetland 2-Ply Spindrift - 450 (500, 525) grams; shown in Seaweed (253) on page 19.
Needles: 16" and 24" circular US 2 (3 mm) and 24" circular US 3 (3.25 mm), *or correct needles to obtain gauge.*
Accessories: Stitch holders.

Measurements
Chest: 45 (49, 53)".
Length to Underarm: 16 (16½, 16½)".
Armhole Depth: 10 (10½, 10½)".
Length: 26 (27, 27)".
Sleeve Length: 20 (21, 21)".

Gauge
On US 3 in **Chart A**: 26 sts and 38 rows = 4".

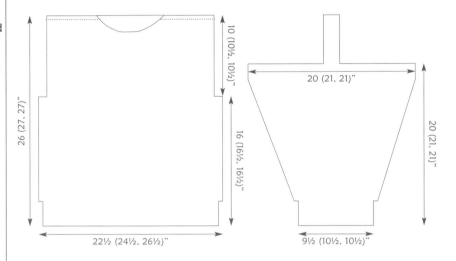

Designer Note
The stitch count of Chart C changes from row to row. It begins and ends with 21 stitches. All stitch counts given assume 21 stitches. Your count may differ, depending on which row you are knitting.

Notes on Charts
Work odd-numbered (RS) rows from right to left, and even-numbered (WS) rows from left to right.

Back
With US 2, CO 174 (190, 206) sts.

Foundation Row (WS): ([K2, p2] 6 (8, 10) times), k2; p1, k2, p9, k2, p1; k2, p2, k2, p3, k1, p3, k2, p2, k2; p1, k2, p9, k2, p1; ([k3, p4] 3 times), k3; p1, k2, p9, k2, p1; k2, p2, k2, p3, k1, p3, k2, p2, k2; p1, k2, p9, k2, p1; ([k2, p2] 6 (8, 10) times), k2.

Row 1 (RS): Work ribbing as set over first 26 (34, 42) sts; work Row 1 of **Chart B** over next 15 sts; work ribbing as set over next 19 sts; work Row 1 of **Chart B** over next 15 sts; work Row 1 of **Chart E** over next 24 sts; work Row 1 of **Chart B** over next 15 sts; work ribbing as set over next 19 sts; work Row 1 of **Chart B** over next 15 sts; work ribbing as set over last 26 (34, 42) sts.

Continuing ribbing and charts as set, work until there are 28 rows of ribbing (including Foundation Row).

Next Row (WS) (Increase Row): ([K2, p2] 6 (8, 10) times), k2; work Row 4 of **Chart B**; k3, m1, k3, p3, k1, p3, k3, m1, k3; work Row 4 of **Chart B**; ([k2, m1, k1, p4] 3 times), k1, m1 k2; work Row 4 of **Chart B**; k3, m1, k3, p3, k1, p3, k3, m1, k3; work Row 4 of **Chart B**; ([k2, p2] 6 (8, 10) times), k2 (182 (198, 214) sts on needle).

Change to US 3.

Next Row (RS): Work Row 1 of **Chart A** over first 26 (34, 42) sts; work Row 5 of **Chart B**; work Row 1 of **Chart C**; work Row 5 of **Chart B**; work Row 1 of **Chart D**; work Row

duxbury point pullover

5 of **Chart B**; work Row 1 of **Chart C**; work Row 5 of **Chart B**; work Row 1 of **Chart A** over last 26 (34, 42) sts.

Continuing charts as set, work until piece measures 16 (16½, 16½)" from CO edge, ending with RS facing for next row.

Shape Armholes
BO 8 sts at beg of next 2 rows. Continue as set without further shaping on rem 166 (182, 198) sts *(see Designer Note)* until piece measures 25½ (26½, 26½)" from CO edge; end by working Row 8 (16, 16) of Chart D (RS faces for next row).

Back Neck and Shoulders
Next Row (RS): Knit and dec evenly over next 52 (60, 68) sts *(see Designer Note)* to equal 34 (39, 45) sts; place these sts on holder for right shoulder strap attachment. Work next 62 sts and place on holder for back neck. Knit and dec evenly over next 52 (60, 68) sts to equal 34 (39, 45) sts; place these sts on holder for left shoulder strap attachment.

Front
Work same as for back until piece measures 22½ (23½, 23½)" from CO edge; end by working Row 16 (8, 8) of Chart D (RS faces for next row).

Shape Front Neck
Next Row (RS): Work 67 (75, 83) sts *(see Designer Note)*; work next 32 sts and place on holder for front neck; work rem 67 (75, 83) sts.

Turn, and working each side separately, dec 1 st at neck edge on every row 15 times. Work without further shaping on rem 52 (60, 68) sts *(see Designer Note)* until piece measures 25½ (26½, 26½)" from CO edge, ending with RS facing for next row.

Next Row (RS): Knit, dec'g same as for back shoulders. Place rem 34 (39, 45) sts *(see Designer Note)* on holders for shoulders.

Sleeves
With US 2, CO 63 (67, 67) sts.

Foundation Row (WS): K0 (2, 2), p2, k2; p1; k2, p2, k2, p3, k1, p3, k2, p2, k2; p1; k2, p2, k2, p1; k2, p9, k2, p1; k2, p2, k2, p3, k1, p3, k2, p2, k2; p1; k2, p2; k0 (2, 2).

Row 1 (RS): Work ribbing as set over first 24 (26, 26) sts; work Row 1 of **Chart B** over next 15 sts; work ribbing as set over next 24 (26, 26) sts.

Continuing ribbing and **Chart B** as set, work until there are 28 rows of ribbing (including **Foundation Row**).

Next Row (WS) (Increase Row): K0 (2, 2), p2, m1, k2; p1; k2, m1, k4, p3, k1, p3, k4, m1, k2; work Row 4 of **Chart B** over next 15 sts; k2, m1, k4, p3, k1, p3, k4, m1, k2; p1; k2, m1, p2, k0 (2, 2) (69 (73, 73) sts on needle).

Change to US 3.

Next Row (RS): Work **Chart A** over next 4 (6, 6) sts; p1, k1tbl; work Row 1 of **Chart C** over next 21 sts; work Row 5 of **Chart B** over next 15 sts; work Row 1 of **Chart C** over next 21 sts; k1tbl, p1; work **Chart A** over next 4 (6, 6) sts.

Continue charts as set, **AND AT SAME TIME**, inc 1 st at beg and end of 2nd row 5 (6, 6) times, every 4th row 13 times, and every 6th row 12 (13, 13) times (129 (137, 137) sts on needle) *(see Designer Note)*. Work without further shaping until sleeve measures 20 (21, 21)" from CO edge, ending with RS facing for next row.

Next Row (RS): BO 55 (59, 59) sts *(see Designer Note)*, work to end of row.

Next Row (WS): BO 55 (59, 59) sts *(see Designer Note)*, work to end of row.

Place rem 19 sts on holder for shoulder strap.

Knit Left Shoulder Strap and Join Left Shoulder
Place sts for left sleeve onto circular US 3, and continuing **Chart B** as set, work first 18 sts of left shoulder strap, then slip last st to right-hand needle. Place sts for front left shoulder onto left-hand needle with RS facing. Slip last st of left shoulder strap back to left-hand needle; ssk last st of left shoulder strap with first st of front left shoulder. Turn.

Next Row (WS): With yarn in front, slip the first purl st of eft shoulder strap; continuing **Chart B** as set, work the next 17 sts of left shoulder strap; slip the last left shoulder strap st to right-hand needle. Place sts for back left shoulder onto left-hand needle with WS facing. Slip last st of left shoulder strap back to left-hand needle. Purl tog last st of left shoulder strap with first st of back left shoulder.

Next Row (RS): Slip first st of left shoulder strap knitwise; work next 17 sts of left shoulder strap; ssk last st of left shoulder strap with next st of front left shoulder.

Next Row (WS): With yarn in front, slip first purl st of left shoulder strap; knit next 17 sts of left shoulder strap; purl tog last st of left shoulder strap with next st of back left shoulder.

duxbury point pullover

Rep these last two rows until all left front and left back shoulder sts have been joined with left shoulder strap. Place 19 shoulder strap sts on holder.

Knit Right Shoulder Strap and Join Right Shoulder
Rep for right sleeve, substituting left shoulder strap with right shoulder strap and left front shoulder with right front shoulder and left back shoulder with right back shoulder.

Sew tops of sleeves to armholes.

Neckband
With 16" US 2, beg at right shoulder strap, k3tog, p2, k9 (work cable crossing if necessary), p2, k3tog; work 62 (62, 62) sts from back neck holder as follows:

1st Size Only (Back Neck Sts): P2, k1, p2, k2tog, k1, p2tog, p1, k2tog, k1, p2, k1; ([p2, k2] twice), p1, p2tog, p1, k1, k2tog, k1, p1, p2tog, p1, ([k2, p2] twice); k1, p2, k1, k2tog, p1, p2tog, k1, k2tog, p2, k1, p2.

2nd & 3rd Sizes Only (Back Neck Sts): P2, k1, p2, k2tog, k1, p2tog, p1, k2tog, k1, p2, k1; ([p1, p2tog, p1, k1, k2tog, k1] 3 times), p1, p2tog, p1; k1, p2, k2tog, k1, p2tog, p1, k2tog, k1, p2, k1, p2.

All Sizes: Work left shoulder strap as follows: k3tog, p2, k9 (work cable crossing if necessary), p2, k3tog; pick up 19 (21, 21) sts down left neck edge; work 32 (32, 32) sts from front neck holder as follows:

1st Size Only (Front Neck Sts): K2tog, ([p1, p2tog, p1, k1, k2tog, k1] 3 times), p1, p2tog, p1, k2tog.

2nd & 3rd Sizes Only (Front Neck Sts): P1, k1, ([p2, k2] twice), p1, p2tog, p1, k1, k2tog, k1, p1, p2tog, p1, ([p2, k2] twice), k1, p1.

All Sizes: Pick up 19 (21, 21) sts up right neck edge.

Place marker to indicate beg of rnd.

Next 11 Rnds: Continue right shoulder strap sts in **Chart B** as set; work back neck sts in rib as set; work left shoulder strap sts in **Chart B** as set; work left neck sts as follows:

1st Size Only: ([P2, k2] 4 times), p2, k1; work front neck sts in rib as set; work right neck sts as follows: k1, ([p2, k2] 4 times), p2.

2nd & 3rd Sizes Only: ([P2, k2] 5 times), p1; work front neck sts in rib as set; work right neck sts as follows: p1, ([k2, p2] 5 times).

BO in pattern.

Finishing
Sew side and sleeve seams. Weave in ends. Block gently.

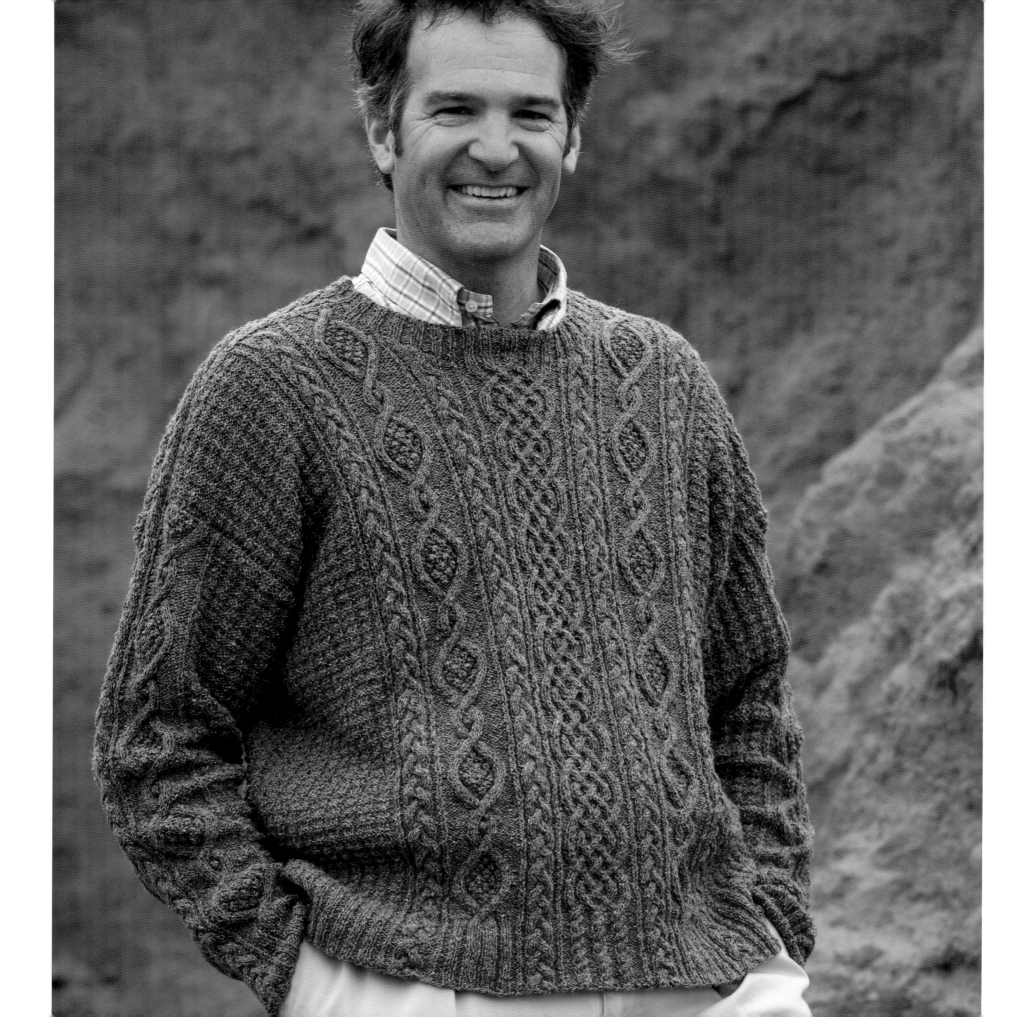

duxbury point pullover

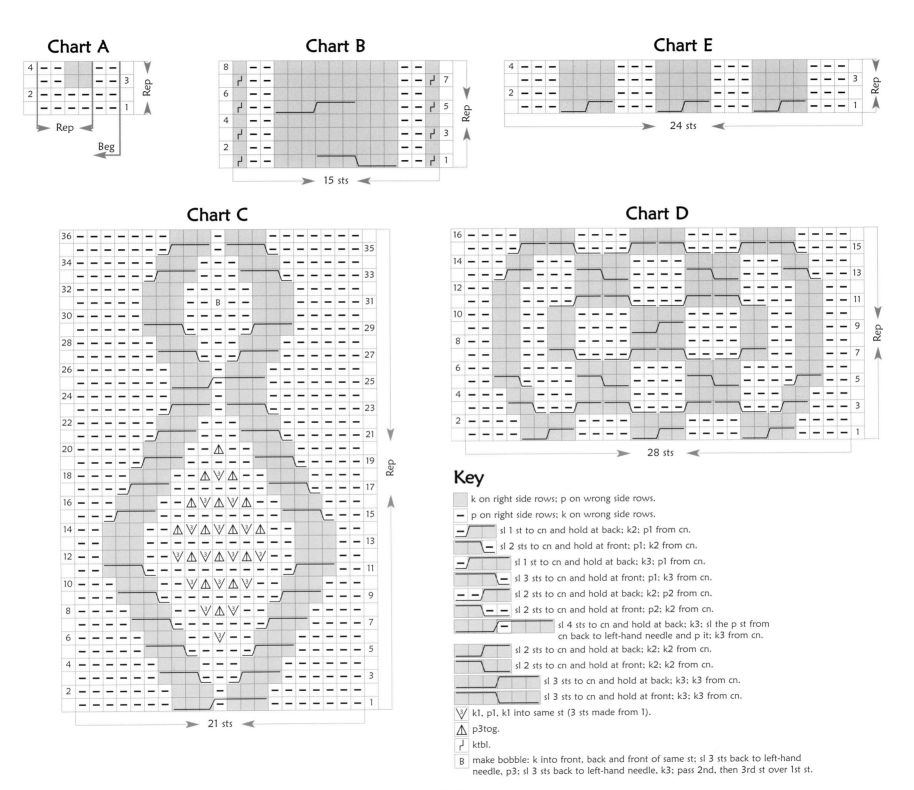

duxbury point wrap

beatrice smith

MATERIALS
YARN: Jamieson's Shetland 2-Ply Spindrift - 275 grams. Shown in Sand (183) on facing page and on page 5.
NEEDLES: US 3 (3.25 mm), *or correct needles to obtain gauge.*

MEASUREMENTS
LENGTH: Approx. 55" (not including tassels).

GAUGE
On US 3 in **Chart A**: 26 sts and 38 rows = 4".

NOTE
Use same charts as for **Duxbury Point Pullover**, on page 21.

WRAP
CO 134 sts.

Foundation Row (WS): K2, p2, k2; p1, k2, p9, k2, p1; k2, p2, k2, p3, k1, p3, k2, p2, k2; p1, k2, p9, k2, p1; ([k3, p4] 3 times)], k3; p1, k2, p9, k2, p1; k2, p2, k2, p3, k1, p3, k2, p2, k2; p1, k2, p9, k2, p1; k2, p2, k2.

Row 1 (RS): Work ribbing as established over first 6 sts; work Row 1 of **Chart B** over next 15 sts; work ribbing as established over next 19 sts; work Row 1 of **Chart B** over next 15 sts; work Row 1 of **Chart E** over next 24 sts; work Row 1 of **Chart B** over next 15 sts; work ribbing as established over next 19 sts; work Row 1 of **Chart B** over next 15 sts; work ribbing as established over last 6 sts.

Continuing ribbing and charts as set until there are 8 rows (including Foundation Row).

Next Row (WS) (Increase Row): K2, p2, k2; work Row 8 of **Chart B**; k3, m1, k3, p3, k1, p3, k3, m1, k3; work Row 8 of **Chart B**; ([k2, m1, k1, p4] 3 times), k1, m1 k2; work Row 8 of **Chart B**; k3, m1, k3, p3, k1, p3, k3, m1, k3; work Row 8 of **Chart B**; k2, p2, k2 (142 sts on needle).

Next Row (RS): Work Row 1 of **Chart A** over first 6 sts; work Row 1 of **Chart B**; work Row 1 of **Chart C**; work Row 1 of **Chart B**; work Row 1 of **Chart D**; work Row 1 of **Chart B**; work Row 1 of **Chart C**; work Row 1 of **Chart B**; work Row 1 of **Chart A** over last 6 sts.

Continuing all charts as set, work through Row 15 of the 33rd rep of **Chart D** and Row 23 of the 15th rep of **Chart C**.

Next Row (WS) (Decrease Row): K2, p2, k2; work **Chart B** as set; k2, p1, p2tog, k2, p3, k1, p3, k2, p2tog, p1, k2; work **Chart B** as set; ([k2, k2tog, p4] 3 times), k2tog, k2; work **Chart B** as set; k2, p1, p2tog, k2, p3, k1, p3, k2, p2tog, p1, k2; work **Chart B** as set; k2, p2, k2 (134 sts on needle).

Next Row (RS): Work ribbing as set over first 6 sts; work **Chart B** as set; work ribbing as set over next 19 sts, working cable crossing (Row 25 of **Chart C**) in center 7 sts; work **Chart B** as set set; work Row 1 of **Chart E**; work **Chart B** as set; work ribbing as set over next 19 sts, working cable crossing (Row 25 of **Chart C**) in center 7 sts; work **Chart B** as set; work ribbing as set over last 6 sts.

Continue ribbing and charts as set for 7 more rows (*Chart C continues in ribbing as set*); BO on Row 8 of **Chart E**.

FINISHING
Block gently to finished measurements. If desired, make tassels and attach to short ends.

passive polka dots

carol lapin

MATERIALS
Yarn: Jamieson's 2-Ply Shetland Spindrift - 250 grams. Shown in Black (999).
Needles: 16" and 24" circular US 4 (3.50 mm) *or correct needle to obtain gauge*.

MEASUREMENTS
Before Blocking—
Chest: 38". Length: 16". Sleeve Length: 13½". Sleeve at Top: 19". Sleeve at Cuff: 15".
After Blocking—
Chest: 38". Length: 20". Sleeve Length: 17". Sleeve at Top: 19". Sleeve at Cuff: 15".

GAUGE (before blocking)
On US 3 in garter st: 25 sts and 39 rows = 4".

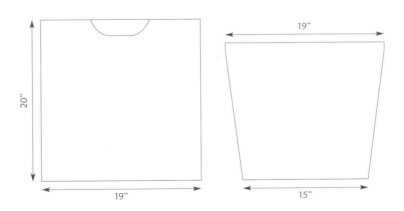

DESIGNER NOTES
Knit body in one piece beginning at bottom of front and working over shoulders to bottom of back. Knit sleeves separately. Block body and sleeves before sewing together.

STITCH PATTERN
Rows 1-12: Knit.
Row 13 (RS): K4, ([BO 13, k11] 4 times), BO 13, k4.
Rows 14 (WS): K4, ([CO 13, k11] 4 times), end CO 13, k4.
Rows 15-24: Knit.
Row 25 (RS): K16, ([BO 13, k11] 3 times), BO 13, k16.
Row 26 (WS): K16, ([CO 13, k11] 3 times), CO 13, k16.

Rep Rows 1-26.

BODY (Knitted in one piece)
CO 117 sts. Work Rows 1-26 of **Stitch Pattern** 5 times, then work Rows 1-13 once.

SHAPE RIGHT NECK
Next Row (Row 14) (WS): K4, CO 13, k11, CO 13, k11 (52 sts). Place rem 26 sts on holder to work left neck shaping later.

Next Row (Row 15) (RS): BO 3 (neck edge); work to end of row (49 sts rem).
Next Row (Row 16): Knit.
Next Row (Row 17): BO 3; work to end of row (46 sts rem).

passive polka dots

Next Row (Row 18): Knit.
Next Row (Row 19): BO 3; work to end of row (43 sts rem).
Next Row (Row 20): Knit.
Next Row (Row 21): BO 2; work to end of row (41 sts rem).
Next Row (Row 22): Knit.
Next Row (Row 23): BO 2; work to end of row (39 sts rem).
Next Row (Row 24): Knit.
Next Row (Row 25): BO 2; k8, BO 13, k16 (37 sts rem).
Next Row (Row 26): K16, CO 13, k8.
Next Row (Row 1): BO 2; work to end of row (35 sts rem).
Next Row (Row 2): Knit.
Next Row (Row 3): BO 2; work to end of row (33 sts rem).
Next 3 Rows (Rows 4-6): Knit.

Leave sts on needle (you'll work across entire row after you've finished shaping left neck).

Shape Left Neck
Reattach yarn at neck edge and shape left neck as follows:

Next Row (Row 14) (WS): BO 3 (neck edge), work 8, CO 13, k11, CO 13, k4 (49 sts on needle.)
Next Row (Row 15) (RS): Knit.
Next Row (Row 16): BO 3; work to end of row (46 sts rem).
Next Row (Row 17): Knit.
Next Row (Row 18): BO 3; work to end of row (43 sts rem).
Next Row (Row 19): Knit.
Next Row (Row 20): BO 2; work to end of row (41 sts rem).
Next Row (Row 21): Knit.
Next Row (Row 22): BO 2; work to end of row (39 sts rem).
Next Row (Row 23): Knit.
Next Row (Row 24): BO 2; work to end of row (37 sts rem).
Next Row (Row 25): K16, BO 13, work 8.
Next Row (Row 26): BO 2; work 6, CO 13, work 16 (35 sts rem).
Next Row (Row 1): Knit.
Next Row (Row 2): BO 2; work to end (33 sts rem).
Next 4 Rows (Rows 4-6): Knit.

On the next row, you'll work across all sts on needle (including those left on needle from right neck shaping), as follows, thus making the back neck edge:

Next Row (Row 7): K33, CO 51, k33 (117 sts on needle).

Work Rows 8-26 of **Stitch Pattern**, then work Rows 1-26 five times. Knit 12 rows. BO loosely.

Neck
With 16" circular needle, RS facing, beg at right shoulder, pick up 48 sts along back neck edge, 20 sts down left front neck edge, 11 sts along front neck edge, and 20 sts up right front neck edge (99 sts on needle). Join, place marker for beg of rnd. Knit 4 rows. BO loosely.

Sleeves
Knit the sleeves from the top down. Isolate the pattern with markers at beg and end of row. You'll dec the 12 sts away at each edge as you work down the sleeve towards the cuff.

CO 117 sts.

Rows 1-12: Knit.
Row 13 (RS): K12, place marker, k4, ([BO 13, k11] 3 times), end BO 13, k4, place marker, k12.
Row 14 (WS): K12, slip marker, k4, ([CO 13, k11] 3 times), end with CO 13, k4, slip marker, k12.

Work Rows 15-26 of **Stitch Pattern**, then work Rows 1-26 four more times (keeping first 12 and last 12 sts in garter st), **AND AT SAME TIME**, dec 1 st at beg and end of every 4th row until you have decreased all 12 sts set off by markers at edges. Work Rows 1-14 of **Stitch Pattern**. Knit 8 rows. BO loosely.

Finishing
Weave in ends. Block pieces to finished measurements, then sew sleeves to body. Sew side and sleeve seams.

inverness cape

carol lapin

MATERIALS
Yarn: Jamieson's Shetland Heather Aran - 800 grams. Shown in Conifer (336)
Needles: US 6 (4 mm) and 40" circular US 8 (5 mm), *or correct needles to obtain gauge*.
Accessories: Three 7/8" buttons.

MEASUREMENTS
Circumference at Widest Point: 68".
Length (including bottom ruffle): 27".

GAUGE
On US 8 in **Seed Stitch**: 16 sts and 28 rows = 4"

Designer Note
Work cape in one piece from the neck downward.

Seed Stitch (Even No. of Sts)
Row 1 (RS): *K1, p1; rep from *.
Row 2 (WS): Purl the knit sts and knit the purl sts as they face you.

Rep Rows 1-2.

Neck
With US 8, CO 160 sts. Beg with RS row, work in **Seed Stitch** for 9 rows. Change to US 6.

Next Row (WS): *K2tog, p2tog; rep from * (80 sts rem).
Next Row (RS): *K1, p1; rep from *.

Next Row (Begin Buttonhole) (WS): Continuing in ribbing as set, work to last 5 sts; BO 3, work 2.
Next Row (Complete Buttonhole) (RS): Work 2, CO 3; work to end of row.

*Continue in ribbing as set for 10 rows, then work the 2 buttonhole rows. Rep from * once (3 buttonholes made). Work 3 more rows in ribbing.

Body
Change to US 8 and **Seed Stitch**.

Next Row (RS): Work 15 sts; place marker; work 10 sts; place marker; work 30 sts; place marker; work 10 sts; place marker; work 15 sts.

Next Row (WS): Work in **Seed Stitch** as set.

Continuing in **Seed Stitch** as set, inc 1 st before and after each marker on every RS row (working inc'd sts into pattern) 15 times (200 sts on needle), then inc 1 st before and after each marker every 4th row 9 times (272 sts on needle). Work without further shaping until body measures 21" (not including neck), ending with RS facing for next row.

Ruffle
Next Row (RS): *K1, yo; rep from * (544 sts on needle).

Work in **Seed Stitch** until ruffle measures 6". BO.

Finishing
Sew on buttons opposite buttonholes. Weave in ends. Block gently to finished measurements.

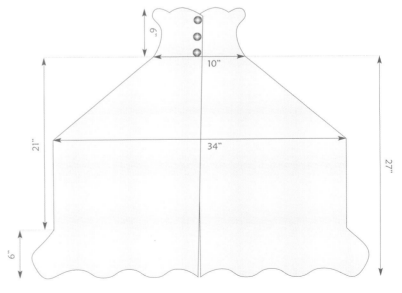

tomales bay skirt

carol lapin

MATERIALS

YARN: Jamieson's Shetland Double Knitting - 25 grams each of Burnt Umber (1190), Foxglove (273), Loganberry (1290), Maroon (595), Mogit (107), Moorgrass (286), Moss (147), Sholmit (103), Sunrise (187) and Thistledown (237); 50 grams each of Black (999), Blue Lovat (232), Bracken (231), Dusk (165), Moorland (195), Pacific (763) and Pine (234); 75 grams of Oxford (123).

NEEDLES: 16" or 20" circular US 4 (3.50 mm) (for waistband) and 32" circular US 6 (4 mm) (for main body of skirt), *or correct needles to obtain gauge.*

MEASUREMENTS

LENGTH (measured at center front including waist & bottom roll): 24".
WAIST: 29" (measured around all 4 points).

GAUGE

On US 6 in st st: 21 sts and 30 rows = 4".
Do not measure gauge on the diagonal.

STRIPE SEQUENCE (WORKED FROM WAIST DOWN, NOT INCLUDING WAISTBAND OR BOTTOM ROLL)

5 rows Moorgrass	2 rows Burnt Umber	1 row Moorland	1 row Maroon
2 rows Loganberry	1 row Moorland	2 rows Burnt Umber	3 rows Moorland
4 rows Pacific	2 rows Sunrise	2 rows Moorland	4 rows Oxford
2 rows Bracken	3 rows Moorland	2 rows Dusk	2 rows Pine
2 rows Oxford	1 row Black	1 row Maroon	3 rows Sholmit
1 row Maroon	1 row Moorgrass	2 rows Pine	1 row Black
2 rows Oxford	4 rows Dusk	2 rows Moorgrass	3 rows Bracken
4 rows Moorland	2 rows Moss	3 rows Sholmit	3 rows Foxglove
3 rows Pine	1 row Bracken	2 rows Pacific	4 rows Blue Lovat
1 row Black	3 rows Mogit	4 rows Moorland	2 rows Dusk
2 rows Burnt Umber	2 rows Pacific	2 rows Bracken	2 rows Sunrise
2 rows Blue Lovat	1 row Foxglove	4 rows Oxford	4 rows Mogit
2 rows Burnt Umber	1 row Loganberry	2 rows Foxglove	2 rows Pine
2 rows Foxglove	1 row Maroon	2 rows Blue Lovat	2 rows Burnt Umber
3 rows Sholmit	3 rows Bracken	4 rows Mogit	2 rows Moorgrass
2 rows Moorgrass	4 rows Oxford	2 rows Sunrise	1 row Loganberry
3 rows Pacific	1 row Thistledown	1 row Black	1 row Maroon
1 row Maroon	1 row Sunrise	3 rows Pine	4 rows Pacific
4 rows Thistledown	4 rows Pine	4 rows Dusk	3 rows Thistledown
2 rows Bracken	3 rows Blue Lovat	2 rows Burnt Umber	1 row Black
1 row Loganberry	1 row Sholmit	2 rows Moorgrass	4 rows Bracken
4 rows Pine	4 rows Thistledown	3 rows Pacific	2 rows Foxglove
2 rows Blue Lovat	2 rows Loganberry	2 rows Loganberry	4 rows Sholmit
4 rows Oxford	1 row Black	3 rows Thistledown	

tomales bay skirt

DESIGNER NOTES
Knit the skirt from the top down, working bottom points in short rows, then sew front and back together. Pick up sts along top to work waistband.

SPECIAL ABBREVIATIONS
RS W&T (wrap & turn for RS rows)—With yarn in back, sl next st as if to purl. Bring yarn to front of work and sl st back to left-hand needle. Turn work.

WS W&T (wrap & turn for WS rows)—With yarn in front, slip next st as if to purl. Bring yarn to back of work and sl st back to left-hand needle. Turn work.

SKIRT FRONT
With US 6 and Moorgrass, CO 77 sts. Following **Stripe Sequence** throughout, work as follows:

Row 1 (RS): Knit to center st, [k1, yo, k1] in center st, knit to end of row.
Row 2 (WS): Purl.

Rep Rows 1-2 until there are 209 sts on needle (104 sts on each side of center st), ending with RS facing for next row.

WORK BOTTOM POINTS
Continuing **Stripe Sequence**, work as follows:

Row 1 (RS): K103, RS W&T.
Row 2 (WS): Purl.
Row 3 (RS): Knit to 2 sts before previous wrap, RS W&T.

Rep Rows 2-3 until there are 15 sts in your short row.

Next Row (WS): Change to Oxford and p15 sts.
Next Row (RS): Knit to center st, knitting wrap loop tog with st it wraps along row. Break yarn.

Sl center st and following 104 sts onto right-hand needle. Turn.

Next Row (WS): P103, WS W&T.
Next Row (RS): Knit.
Next Row (WS): Purl to 2 sts before previous wrap, WS W&T.

Rep last 2 rows until there are 15 sts in your short row.

Next Row (RS): With Oxford, k15 sts.
Next Row (WS): Purl to center st, purling together wrap and st it wraps along row. Break yarn.

Sl sts on left-hand needle to right-hand needle. Turn. Continuing with Oxford, work as follows:

Next Row (RS): Knit.
Next Row (WS): Purl.
Next Row (RS): Knit.
Next Row (WS): Purl.

BO.

SKIRT BACK
Work same as for skirt front.

FINISHING
Sew front and back together along side edges.

WAISTBAND
With US 6, pick up 1 st in each st along top of skirt (approx. 150 sts on needle; be sure number is an even number). Change to circular US 4, place marker and work in the rnd as follows:

Next 18 Rnds: *K1, p1; rep from *.

BO loosely. Weave in ends. Block gently to finished measurements. Oxford at bottom should roll up and hide the short row shaping along edge.

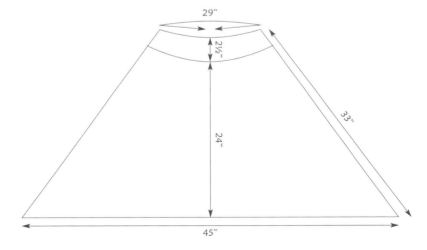

tomales bay cowl

prudence makepeace

MATERIALS
Yarn: Simply Shetland Lambswool & Cashmere - 550 (600, 650, 650, 700) grams. Shown in Velvet (384).
Needles: 24" circular US 6 (4 mm), *or correct needles to obtain gauge*.
Accessories: Stitch holders.

MEASUREMENTS
Chest: 38 (40, 42, 44, 47)".
Length to Armhole: 15 (15, 15½, 15½, 16)".
Armhole Depth: 7 (7, 7½, 7½, 8)".
Length: 22 (22, 23, 23, 24)".
Sleeve (cuff to underarm): 10½ (10½, 10½, 10½, 11)".

On US 6 in st st: 22 sts and 30 rows = 4".

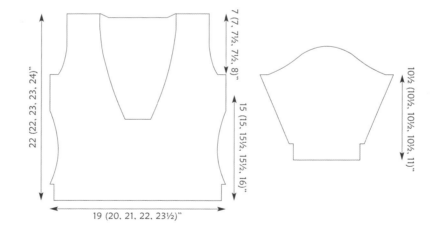

tomales bay cowl

Note
Where only one number is given, it applies to all sizes.

K2 P2 Ribbing
Row 1 (RS): *K2, p2; rep from *; end k2.
Row 2 (WS): Knit the knit sts and purl the purl sts as they face you.

Back
CO 98 (102, 110, 114, 118) sts. Work in **K2 P2 Ribbing** for 2", inc'g 6 (8, 6, 8, 12) sts evenly along last (WS) row (104 (110, 116, 122, 130) sts on needle).

Next Row (RS): Knit.

Continue in st st, **AND AT SAME TIME**, dec 1 st at beg and end of every 4th row 0 (0, 3, 0, 6) times, then every 6th row 4 (4, 2, 4, 0) times (96 (102, 106, 114, 118) sts on needle). Now inc 1 st at beg and end of every 12th (12th, 10th, 12th, 8th) row 4 (4, 5, 4, 6) times (104 (110, 116, 122, 130) sts on needle). Continue without further shaping until piece measures 15 (15, 15½, 15½, 16)" from CO edge.

Shape Armholes
BO 5 (5, 5, 6, 6) sts at beg of next 2 rows, 2 sts at beg of next 4 (4, 4, 6, 6) rows, then dec 1 st at beg and end of every other row 3 (5, 6, 5, 7) times (80 (82, 86, 88, 92) sts on needle). Continue without further shaping until work measures 21½ (21½, 22½, 22½, 23½)" from CO edge, ending with RS facing for next row.

Shape Back Neck
Next Row (RS): Work 25 (26, 27, 28, 29) sts, BO next 30 (30, 32, 32, 34) sts for back neck, work rem 25 (26, 27, 28, 29) sts.

Turn, and working each side separately, BO off at neck edge 3 sts once and 2 sts once. Place rem 20 (21, 22, 23, 24) sts on holders for shoulders.

Front
Work as for back until piece measures 9 (9½, 10, 10¼, 11)" from CO edge, ending with RS facing for next row.

Shape Front Neck
Next Row (RS): Continuing waist shaping as for back, work up to, then BO center 14 (16, 18, 20, 22) sts, work rem sts.

Continuing waist shaping as for back, turn, and working each side separately, dec 1 st at neck edge on 2nd row 1 time, every 6th row 11 (3, 3, 2, 2) times, and every 8th row 1 (8, 8, 8, 8) times. **AT SAME TIME**, when work measures 15 (15, 15½, 15½, 16)" from CO edge, shape armhole same as for back. When piece measures 22 (22, 23, 23, 24)" from CO edge, place rem 20 (21, 22, 23, 24) sts on holders for shoulders.

Sleeves
CO 46 (46, 50, 50, 54) sts. Work in **K2, P2 Ribbing** for 2", inc'g 2 (4, 2, 4, 2) sts evenly along last (WS) row (48 (50, 52, 54, 56) sts on needle).

Change to st st, **AND AT SAME TIME**, inc 1 st at beg and end of every 2nd row 3 (7, 7, 11, 12) times, then every 4th row 14 (12, 12, 10, 10) times (82 (88, 90, 96, 100) sts on needle). Continue without further shaping until piece measures 10½ (10½, 10½, 10½, 11)" from CO edge.

Shape Sleeve Cap
BO 5 (5, 5, 6, 6) sts at beg of next 2 rows, 2 sts at beg of next 4 (4, 4, 6, 6) rows, then decrease 1 st at beg and end of every other row 11 (10, 13, 15, 13) times, then BO 2 sts at beg of next 10 (14, 12, 8, 12) rows, then 3 sts at beg of next 2 rows. BO rem 16 (16, 16, 20, 20) sts.

Join Shoulders
Join shoulders using 3-needle bind-off method.

Cowl Neck
With RS facing, beg at left shoulder seam, pick up 85 (84, 84, 83, 83) sts down left front neck edge, 14 (16, 18, 20, 22) sts along front neck edge, 85 (84, 84, 83, 83) sts up right front neck edge, and 40 (40, 42, 42, 44) sts along back neck edge (224 (224, 228, 228, 232) sts on needle). Join and work in the rnd as follows:

Every Rnd: *K2, p2; rep from *.

When neck measures 8½", BO loosely.

Finishing
Sew sleeves into armholes. Sew side and sleeve seams. Weave in ends. Block gently to finished measurements.

dillon beach pullover

gregory courtney

MATERIALS

YARN: Simply Shetland Silk & Lambswool - 550 (650, 750) grams. Hold 2 strands together throughout. Shown in Ethie (010).
NEEDLES: US 5 (3.75 mm) and US 7 (4.5 mm), *or correct needles to obtain gauge*. 16" circular US 5 (3.75 mm).

MEASUREMENTS

CHEST: 37 (46, 54)".   **LENGTH TO UNDERARM:** 14½ (15, 15½)".   **ARMHOLE DEPTH:** 9½ (10, 10½)".
LENGTH: 24 (25, 26)".   **SLEEVE LENGTH:** 18 (19, 20)".

GAUGE

On US 7 in **Body Chart** or **Sleeve Chart**: 22 sts and 28 rows = 4".

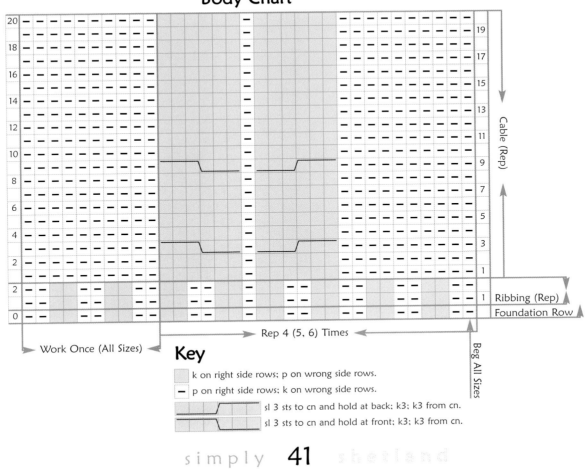

dillon beach pullover

Notes on Charts
Work odd-numbered (RS) rows from right to left, and even-numbered (WS) rows from left to right.

Designer Note
Hold 2 strands of yarn together throughout pattern. Where only one number is given, it applies to all sizes.

Back
With US 5, CO 102 (125, 148) sts. Work **Foundation Row** of **Chart** once, then rep the 2 rows of **Ribbing** until piece measures 2½" from CO edge, ending with RS facing for next row. Change to US 7 and work the 20-row cable rep until piece measures 14½ (15, 15½)" from CO edge, ending with RS facing for next row.

Shape Underarms
BO 5 sts at beg of next 2 rows, then continue as set on rem 92 (115, 138) sts until piece measures 23½ (24½, 25½)" from CO edge, ending with RS facing for next row.

Shape Back Neck
Next Row (RS): Work 28 (39, 49) sts, work next 36 (37, 40) sts and place on holder for back neck, work rem 28 (39, 49) sts.

Turn, and working each side separately, dec 1 st at neck edge 1 time. Work without further shaping on rem 27 (38, 48) sts until piece measures 24 (25, 26)" from CO edge. Place shoulder sts on holders.

Front
Work same as for back until piece measures 20½ (21½, 22½)" from CO edge, ending with RS facing for next row.

Shape Front Neck
Next Row (RS): Work 38 (49, 59) sts, work next 16 (17, 20) sts and place on holder for front neck, work rem 38 (49, 59) sts.

Turn, and working each side separately, dec 1 st at neck edge on every other row 11 times. Work without further shaping on rem 27 (38, 48) sts until piece measures same length as back.

Join Shoulders
Join shoulders using 3-needle bind-off method.

Sleeves
Because increased area is worked in reverse st st, the st gauge for sleeve after cuff is calculated at approx. 19 sts = 4"

With US 5, CO 44 sts. Work **Foundation Row** of **Sleeve Chart** once, then rep the 2 rows of **Ribbing** until piece measures 2½" from CO edge, ending with RS facing for next row. Change to US 7 and work the 20-row cable rep, **AND AT SAME TIME**, while working inc'd sts in reverse st st, inc at beg and end of 1st row 1 time, every 2nd row 4 times, every 4th row 11 (13, 19) times, and every 6th row 7 (7, 4) times (90 (94, 100) sts on needle). Continue without further

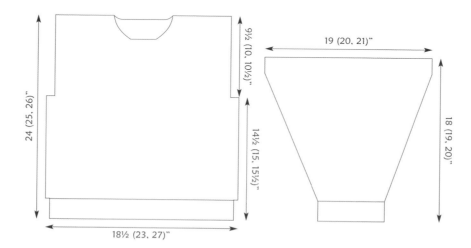

dillon beach pullover

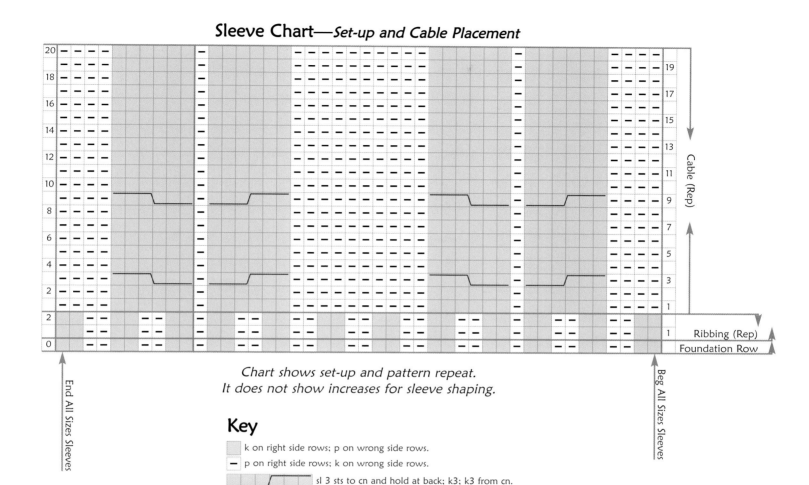

shaping until sleeve measures 18 (19, 20)" from CO edge. BO.

NECKBAND
With 16" circular US 5, beg at right shoulder seam, pick up 4 sts to back neck holder, k36 ([*k19, m1, k18=38 for middle size*], k40) sts from back neck holder, pick up 4 sts to left shoulder seam, pick up 24 sts down left neck edge, k16 ([*k9, m1, k8=18 for middle size*], k20) sts from front neck holder, pick up 24 sts up right neck edge to shoulder seam (108 (112, 116) sts on needle). Join and work in the rnd as follows:
1st & 3rd Sizes Only—Every Rnd: K1, *p2, k2; rep from *; end k1.

2nd Size Only—Every Rnd: *K2, p2; rep from *.

When neckband measures 1¾", BO in pattern.

FINISHING
Sew arms to armholes. Sew side and sleeve seams. Weave in ends. Block to finished measurements.

alamere falls pullover

gregory courtney

MATERIALS
Yarn: Jamieson's Shetland Heather Aran - 550 (600, 650, 700, 750, 800) grams. Shown in Autumn (Hairst) (998).
Needles: 16" circular US 4 (3.50 mm) and circular or straight US 6 (4 mm), *or correct needles to obtain gauge*.
Accessories: Stitch holders.

MEASUREMENTS
Chest: 36 (40, 44, 48, 52, 56)".
Length to Underarm: 14½ (15, 15½, 16, 16½, 17)".
Armhole Depth: 8½ (9, 9½, 10, 10½, 11)".
Length: 23 (24, 25, 26, 27, 28)".
Sleeve Length: 18 (19, 20, 21, 22, 23)".

GAUGE
On US 6 in **Wavy Rib Pattern**: 20 sts and 28 rows = 4".

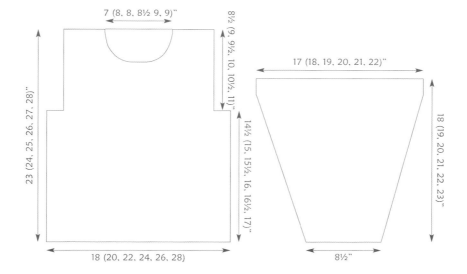

alamere falls pullover

NOTE
Where only one number is given, it applies to all sizes.

RIB PATTERN (MULTIPLE OF 3)
Row 1 (WS): *K1, p2; rep from *.
Row 2 (RS): *K2, p1; rep from *.
Row 3 (WS): *P2, k1; rep from *.
Row 4 (RS): *P1, k2; rep from *.

Rep Rows 1-4.

BACK
With US 4, CO 90 (99, 111, 120, 129, 141) sts. Work in **Rib Pattern** for 2½". Change to US 6 and continue in pattern as set until piece measures 14½ (15, 15½, 16, 16½, 17)" from CO edge, ending with RS facing for next row.

SHAPE UNDERARMS
BO 9 sts at beg of next 2 rows, then continue in pattern as set on rem 72 (81, 93, 102, 111, 123) sts until armhole measures 8½ (9, 9½, 10, 10½, 11)", ending with RS facing for next row.

SHAPE BACK NECK
Next Row (RS): Work 18 (21, 27, 30, 33, 39) sts and place on holder for right shoulder; work 36 (39, 39, 42, 45, 45) sts and place on holder for back neck; work rem 18 (21, 27, 30, 33, 39) sts and place on holder for left shoulder.

FRONT
Work same as for back until armhole measures 4½ (5, 5½, 6, 6½, 7)", ending with RS facing for next row.

SHAPE FRONT NECK
Next Row (RS): Work 30 (33, 39, 42, 45, 51) sts; work 12 (15, 15, 18, 21, 21) sts and place on holder for front neck; work rem 30 (33, 39, 42, 45, 51) sts.

Turn, and working each side separately, dec 1 st at neck edge on every WS row 12 times, then work without further shaping on rem 18 (21, 27, 30, 33, 39) sts until front measures same length as back. Place sts on holders for shoulders.

JOIN SHOULDERS
Join shoulders using 3-needle bind-off method.

SLEEVES
Note: Inc's beg on 4th row (RS) from CO edge.

Hint: Work incs on RS rows. When you beg incs, change the first 2 and last 2 sts to st st and inc in pattern after these first 2 and before these last 2 sts. Your rib pattern will grow attractively from the edges, and you'll have a tidy edge when you sew the sleeve seam.

With US 4, CO 42 sts. Work in **Rib Pattern** with US 4 for 2½" then with US 6 thereafter, **AND AT SAME TIME**, inc 1 st at beg and end of every 4th row 7 (13, 15, 21, 23, 29) times, then every 6th row 14 (11, 11, 8, 8, 5) times (84 (90, 94, 100, 104, 110) sts on needle). Work without further shaping until sleeve measures 18 (19, 20, 21, 22, 23)" from CO edge. BO.

NECKBAND
With 16" circular US 4, RS facing, beg at right shoulder seam, k36 (39, 39, 42, 45, 45) sts from back neck holder, pick up 27 sts over shoulder and down left neck edge, k12 (15, 15, 18, 21, 21) sts from front neck holder, pick up 27 sts up right neck edge and over shoulder (102 (108, 108, 114, 120, 120) sts on needle). *It doesn't matter that you pick up exactly the number of sts indicated, so long as it is an even number.* Join, and work in the rnd as follows:

Rnds 1 & 2: Purl.
Rnds 3 & 4: *K1, p1; rep from *.
Rnds 5 & 6: *P1, K1; rep from *.

Rep Rnds 3-6 until neckband measures 1½". BO.

FINISHING
Sew sleeves into armholes. Sew side and sleeve seams. Weave in ends. Block gently.

box stripe pullover

beatrice smith

MATERIALS
Yarn: Jamieson's Shetland Double Knitting - 125 grams each of Coffee (880), Mulberry (598), Old Gold (429), Port Wine (293) and Prussian Blue (726).
Needles: 16" and 24" circular US 5 (3.75 mm) and 47" circular US 6 (4 mm), *or correct needle to obtain gauge*.

MEASUREMENTS
Chest: 54".   **Length (including waistband):** 27".   **Sleeve Length (including cuff):** 16¾".

GAUGE
On US 6 in st st: 21 sts and 28 rows = 4".

ABOUT CHARTS
Work odd-numbered (RS) rows from right to left, and even-numbered (WS) rows from left to right. All charts show front of garment to shoulder line. Therefore, work twice the number of sts shown to include the back. On sleeve charts, work incs (Chart A) and decs (Chart D) on both right and left edges of sleeve.

PULLOVER (Worked Cuff-to-Cuff in One Piece)
With Prussian Blue and US 6, CO 64 sts. Work **Chart A** (inc'g to shape sleeve, as shown) through Row 100.

Row 101 (last row of Chart A) (RS): Work to end of row; CO 75 sts.
Row 102 (first row of Chart B) (WS): Work 75 sts with Coffee only, add Old Gold and work pattern across sleeve portion, break off Old Gold, and with Coffee, CO 75 sts.
Row 103 (RS): Rejoin Old Gold, and work pattern.

Continue **Chart B** as set through Row 165.

Row 166 (WS): Work 132 sts and place on holder for back. BO next 10 sts, work to end of row.

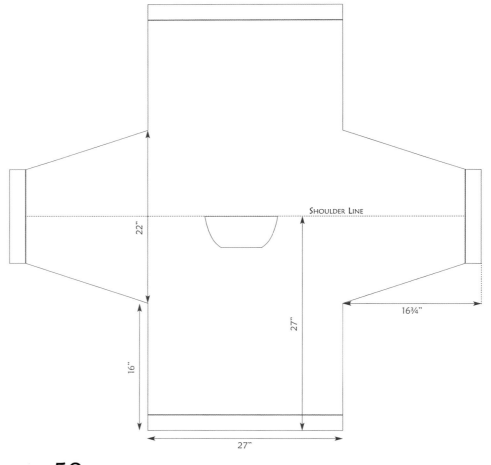

box stripe pullover

Continue as set, shaping neck as shown through Row 196, then work **Chart C**, reversing all shaping for left side of neck. When 10 sts are CO for left neck edge, place front sts on holder and return to back sts. Work back sts until even with front sts, then work both front and back sts together. Work remainder of chart as set, binding off 75 sts twice at left side seam, then continue into **Chart D** (dec'g to shape sleeve, as shown). BO on last row of chart.

NECKBAND
With 16" US 5 and Prussian Blue, RS facing, beg at right shoulder, pick up 110 sts evenly around neck edge. Join, and work in the rnd as follows:

Rnd 1: With Prussian Blue, purl.
Rnd 2: With Mulberry, knit.
Rnd 3: With Mulberry, purl.
Rnd 4: With Old Gold, knit.
Rnd 5: With Old Gold purl.
Rnd 6: With Coffee, knit.
Rnd 7: With Coffee, purl.
Rnd 8: With Port Wine, knit.
Rnd 9: With Port Wine, purl.
Rnd 10: With Prussian Blue, knit.
Rnd 11: With Prussian Blue, purl.
Rnd 12: With Mulberry, knit.

With Mulberry, BO in purl.

RIGHT SLEEVE CUFF
With 16" circular US 5 and Port Wine, RS facing, pick up 54 sts evenly along wrist end of sleeve.

Next Row (WS): With Port Wine, knit.

Continuing in garter st (knit every row), *knit 2 rows each with Coffee, Old Gold, Mulberry, Prussian Blue and Port Wine; rep from * once more; knit 2 rows with Coffee, binding off on 2nd row.

LEFT SLEEVE CUFF
With 16" circular US 5 and Mulberry, pick up 54 sts evenly along wrist end of sleeve.

Next Row (WS): With Mulberry, knit.

Continuing in garter st (knit every row), *knit 2 rows each with Old Gold, Coffee, Port Wine, Prussian Blue and Mulberry; rep from * once more; knit 2 rows with Old Gold, binding off on 2nd row.

WAISTBAND
With 24" circular US 5 and Prussian Blue, RS facing, pick up 134 sts evenly along front bottom edge.

Next Row (WS): With Prussian blue, knit.

Continuing in garter st (knit every row), *knit 2 rows each with Port Wine, Coffee, Old Gold, Mulberry and Prussian Blue; rep from * once more; knit 2 rows with Port Wine, binding off on 2nd row.

Rep waistband instructions for back bottom edge.

FINISHING
Sew side and sleeve seams. Weave in ends. Block to finished measurements.

box stripe pullover

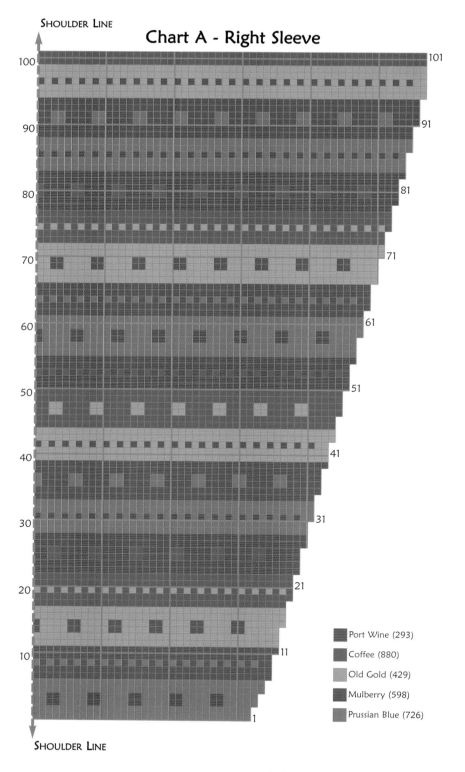

simply 54 shetland

box stripe pullover

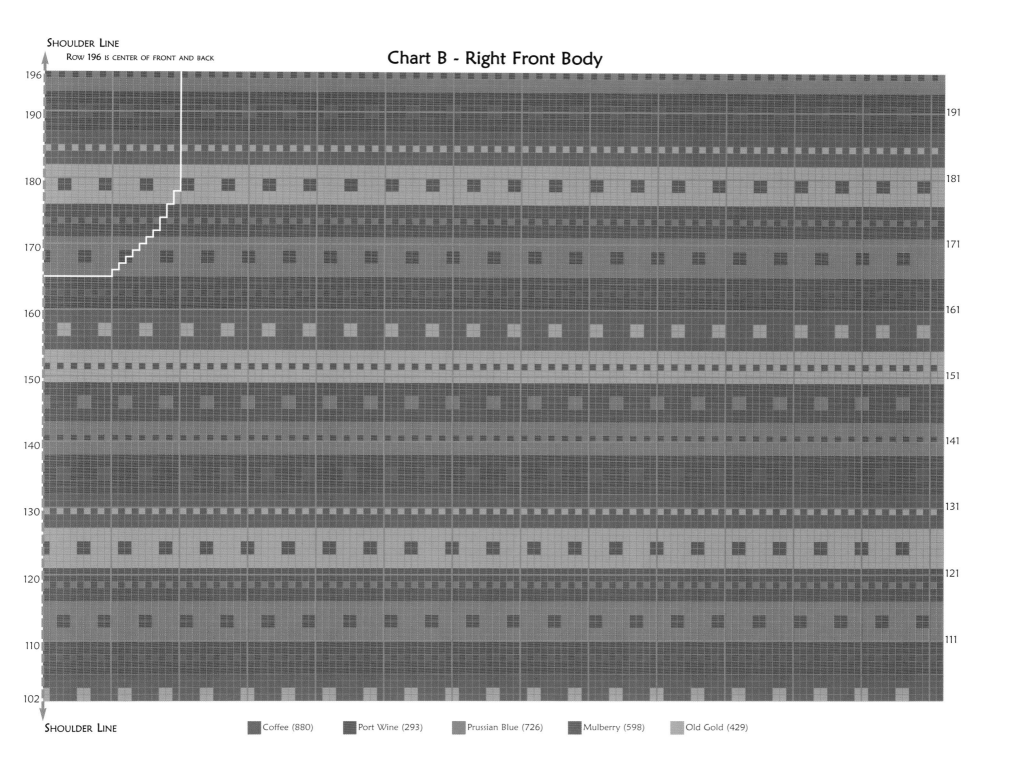

Chart B - Right Front Body

simply 55 shetland

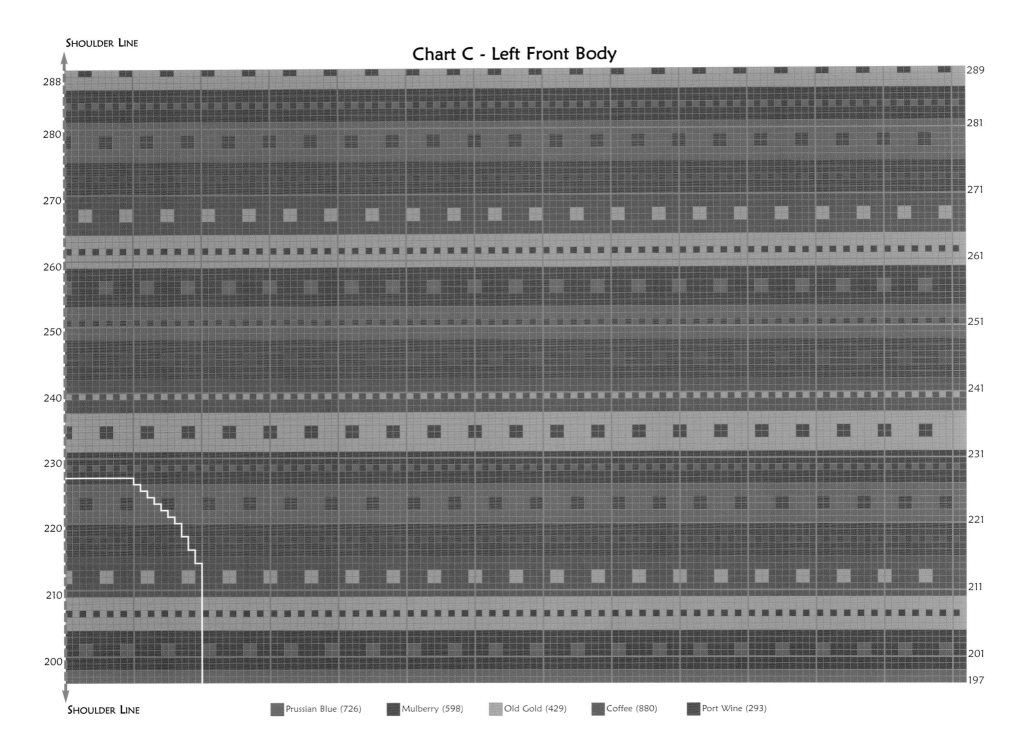

box stripe pullover

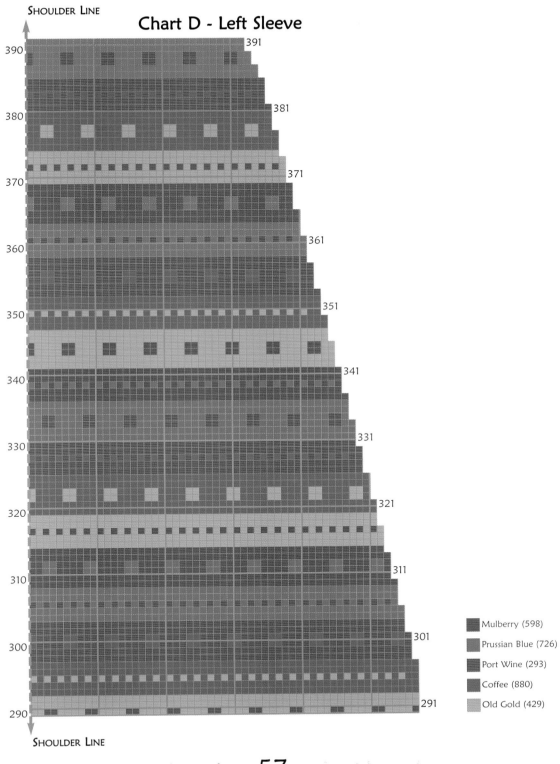

shadow pullover

carol lapin

MATERIALS
YARN: Jamieson's Shetland 2-Ply Spindrift - 250 (275, 275) grams of Color A; 225 (250, 250) grams of Color B; and 25 (25, 25) grams of Color C. Shown in Color A, Oxford (123); Color B, Bracken (231); and Color C, Madder (587).
NEEDLES: 32" and 16" circular US 2 (3 mm), *or correct needle to obtain gauge.*

MEASUREMENTS
CHEST: 43 (48, 53)".
LENGTH: 24 (25, 26)".
SLEEVE LENGTH: 16 (16½, 17)".

GAUGE
On US 2 in **Shadow Stitch for Front & Back**: 24 sts and 48 rows = 4".

NOTE
Where only one number is given, it applies to all sizes.

SHADOW STITCH FOR FRONT & BACK (MULTIPLE OF 16)
Row 1 (RS): With Color A, knit.
Row 2 (WS): With Color A, *k8, p8; rep from *.
Row 3 (RS): With Color B, knit.
Row 4 (WS): With Color B, *p8, k8; rep from *.

Rep Rows 1-4.

SHADOW STITCH FOR SLEEVES
Row 1 (RS): With Color A, knit.
Row 2 (WS): With Color A, k4, ([p8, k8] 3 times), p4.
Row 3 (RS): With Color B, knit.
Row 4 (WS): With Color B, p4, ([k8, p8] 3 times), k4.

Rep Rows 1-4.

BACK
With Color A, CO 128 (144, 160) sts and purl 1 row (WS). Work in **Shadow Stitch for Front & Back** until piece measures 22 (23, 24)" from CO edge, ending with RS facing for next row. Place first 42 (48, 56) sts on holder for right shoulder, next 44 (48, 48) sts on holder for back neck, and rem 42 (48, 56) sts on holder for left shoulder.

FRONT
Work same as for front until piece measures 14½ (15½, 16½)" from CO edge, ending with RS facing for next row.

SHAPE FRONT NECK
Next Row (RS): Work 63 (71, 79) sts, place 2 sts on holder for front neck, work rem 63 (71, 79) sts.

Turn, and working each side separately, dec 1 st at neck edge every 3rd row 21 (23, 23) times. Work without further shaping until piece measures same length as back, ending with RS facing for next row. Place rem 42 (48, 56) shoulder sts on holders.

SLEEVES
With Color A, CO 56 sts and purl 1 row (WS). Work in **Shadow Stitch for Sleeves, AND AT SAME TIME,** inc 1 st at beg and end of 2nd row 1 time, then every following 4th row 20 (20, 22) times, then every 6th row 8 (9, 9) times (114 (116, 120) sts on needle). Continue without further shaping until piece measures 14 (14½, 15)" from CO edge, ending with WS facing for next row.

Next Row (WS): Purl.
BO knitwise in same color as previous row.

JOIN SHOULDERS AND SEW SIDE SEAMS
Join shoulders using 3-needle bind-off method. Center sleeves at shoulder seams and sew to body. Sew side seams,

shadow pullover

leaving sleeve seams unsewn (you'll sew these up after you've knit the cuff).

WAISTBAND
With Color A, RS facing, pick up 224 (252, 280) sts evenly around bottom of sweater. Join and work in the rnd as follows, stranding yarn floats on WS (inside of sweater):

Rnd 1: *P2 Color A, p2 Color C; rep from *.
Rnd 2: *K2 Color A, k2 Color C; rep from *.
Rnd 3: *P2 Color B, p2 Color A; rep from *.
Rnd 4: *K2 Color B, k2 Color A; rep from *.

Rep Rnds 1-4 five times.

Next Rnd: With Color A, purl.

With Color A, BO kwise.

SLEEVE CUFFS
With Color A, RS facing, pick up 54 sts evenly along cuff edge and work as follows:

Row 1 (WS): *K2 Color A, k2 Color C; rep from *; end k2 Color A.
Row 2 (RS): *K2 Color A, k2 Color C; rep from *; end k2 Color A.
Row 3 (WS): *K2 Color B, k2 Color A; rep from *; end k2 Color B.
Row 4 (RS): *K2 Color B, k2 Color A; rep from *; end k2 Color B.

Rep Rows 1-4 four times, ending with WS facing for next row.

Next Row (WS): With Color A, knit.

With Color A, BO kwise.

NECKBAND
With 16" circular needle and Color A, RS facing, k44 (48, 48) sts from back neck holder, pick up 59 sts down left neck edge, place marker, k2 from holder, place marker, pick up 59 sts up right neck edge (164 (168, 168) sts on needle). Join, and work in the rnd as follows:

Rnd 1: P1 Color C, *p2 Color A, p2 Color C; rep from * to 2 sts before marker, k2tog, slip marker, with Color A, k2 center sts, slip marker, k2tog, **p2 Color C, p2 Color A; rep from ** to last st, p1 Color C.

Work 7 more rnds same as for bottom ribbing, **AND AT SAME TIME**, k2tog before 1st marker and after 2nd marker every other rnd.

Next Rnd: With Color A, purl.

With Color A, BO kwise.

FINISHING
Sew sleeve seams. Weave in ends. Block gently to finished measurements.

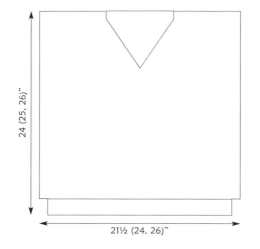
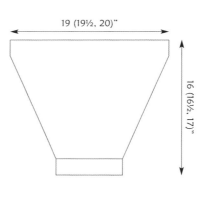

drake's bay jacket & coat

carol lapin

MATERIALS
YARN FOR COAT: In Jamieson's Shetland Heather Aran - 950 (1,000, 1,100, 1,200) grams; shown in Autumn (998) on page 66.
YARN FOR COAT: In Simply Shetland Silk & Lambswool (2 strands held together throughout) - 750 (850, 900, 1,000) grams; not shown.
YARN FOR JACKET: In Simply Shetland Silk & Lambswool (2 strands held together throughout) - 600 (700, 750, 800) grams; shown in Drumlanrig (031) on facing page.
YARN FOR JACKET: In Jamieson's Shetland Heather Aran - 750 (850, 950, 1,000) grams; not shown.
NEEDLES (BOTH JACKET AND COAT): Two 32" circular US 7 (4.5 mm), *or correct needles to obtain gauge*. D/3 crochet hook.
ACCESSORIES: Five 7/8" buttons (coat only); Four 7/8" buttons (jacket only). Stitch holders.

MEASUREMENTS (BOTH JACKET AND COAT)
CHEST: 40½ (44½, 48½, 52½)".
LENGTH: Jacket: 20"; Coat: 39".
SLEEVE LENGTH (cuff to underarm): 12".

GAUGE (BOTH JACKET AND COAT)
On US 7 in st st: 18 sts and 27 rows = 4".
Jamieson's Shetland Heather Aran is used single-strand. Simply Shetland Silk & Lambswool is used double-strand.

SPECIAL ABBREVIATIONS
RS W&T (wrap & turn for RS rows)—With yarn in back, sl next st as if to purl. Bring yarn to front of work and sl st back to left-hand needle. Turn work.

WS W&T (wrap & turn for WS rows)—With yarn in front, slip next st as if to purl. Bring yarn to back of work and sl st back to left-hand needle. Turn work.

SEED STITCH (ANY NO. OF STS)
Row 1 (RS): *K1, p1; rep from * to end of row.
Row 2 (WS): Knit the purl sts and purl the knit sts as they face you.

Rep Rows 1-2.

DESIGNER NOTES
Where only one number is given, it applies to all sizes. Instructions for **Jacket** begin below. Instructions for **Coat** begin on page 64. Both jacket and coat are worked the same from **Divide for Fronts and Back** instructions to end of pattern.

JACKET
BODY (WORKED IN ONE PIECE TO UNDERARMS)
TIER 1
CO 208 (229, 250, 270) sts. Work in **Seed Stitch** for 6 rows, ending with RS facing for next row.

Next Row (RS): Work first 7 sts in **Seed Stitch** as set; k194 (215, 236, 256); work rem 7 sts in **Seed Stitch** as set.

Next row (WS): Work first 7 sts in **Seed Stitch** as set; p194 (215, 236, 256); work rem 7 sts in **Seed Stitch** as set.

Continue as set for 26 more rows, always working first 7 sts and last 7 sts in **Seed Stitch** as set and middle 194 (215, 236, 256) sts in st st as set, ending with RS facing for next row.

MAKE BUTTONHOLE
Next Row (RS): Work 2 sts in **Seed Stitch** as set, BO 3 sts, work 2 sts in **Seed Stitch** as set, continue as set to end of row.
Next Row (WS): Work as set to last 4 sts, work 2 sts in **Seed Stitch** as set, CO 3 sts, work 2 sts in **Seed Stitch** as set.

drake's bay jacket & coat

Hereafter, make buttonholes approx. every 4", making last (4th) buttonhole 2 rows before neck shaping.

Continue as set for 28 more rows, always working first 7 sts and last 7 sts in **Seed Stitch** as set and middle 194 (215, 236, 256) sts in st st as set, ending with RS facing for next row.

Next Row (RS): Work first 7 sts in **Seed Stitch** as set; k7 (5, 10, 11), ([k2tog, k8 (8, 7, 7)] 18 (21, 24, 26) times), k7 (0, 10, 11); work rem 7 sts in **Seed Stitch** as set (190 (208, 226, 244) sts on needle).

Next Row (WS): Work first 7 sts in **Seed Stitch** as set; p176 (194, 212, 230) sts; work rem 7 sts in **Seed Stitch** as set.

Leave sts on needle and set aside.

Tier 2
With another needle, CO 176 (194, 212, 230) sts and work 6 rows in **Seed Stitch**, ending with RS facing for next row.

Join Tier 1 to Tier 2
Next Row (RS): Work first 7 sts of **Tier 1** in **Seed Stitch** as set; place **Tier 2** needle on top of **Tier 1** needle and knit the tiers together; work rem 7 sts of **Tier 1** in **Seed Stitch** as set (190 (208, 226, 244) sts on needle).

Continue **Tier 2** as set for 13 more rows, always working first 7 sts and last 7 sts in **Seed Stitch** as set and middle 176 (194, 212, 230) sts in st st as set, ending with RS facing for next row.

*Proceed to **Divide for Fronts and Back** instructions for coat version. Both jacket and coat versions are worked the same from that point to end of pattern.*

****

COAT
Body (Worked in One Piece to Underarms)
Tier 1
CO 250 (275, 300, 325) sts. Work in **Seed Stitch** for 6 rows, ending with RS facing for next row.

Next Row (RS): Work first 7 sts in **Seed Stitch** as set; k236 (261, 286, 311); work rem 7 sts in **Seed Stitch** as set.

Next Row (WS): Work first 7 sts in **Seed Stitch** as set; p236 (261, 286, 311); work rem 7 sts in **Seed Stitch** as set.

Work last 2 rows until piece measures 9" from CO edge, ending with RS facing for next row.

Next Row (RS): Work first 7 sts in **Seed Stitch** as set; k8 (4, 3, 1), ([k2tog, k9, 9, 8, 8] 20 (23, 28, 31) times), k8 (4, 3, 0); work rem 7 sts in **Seed Stitch** as set (230 (252, 272, 294) sts on needle).

Next Row (WS): Work first 7 sts in **Seed Stitch** as set; p216 (238, 258, 280); work rem 7 sts in **Seed Stitch** as set.

Leave sts on needle.

Tier 2
With another needle, CO 216 (238, 258, 280) sts and work 6 rows in **Seed Stitch**, ending with RS facing for next row.

Join Tier 1 to Tier 2
Next Row (RS): Work first 7 sts of **Tier 1** in **Seed Stitch** as set; place **Tier 2** needle on top of **Tier 1** needle and knit the tiers tog; work rem 7 sts of **Tier 1** in **Seed Stitch** as set (230 (252, 272, 294) sts on needle).

Continue on **2nd Tier** as set for 55 more rows, always working first 7 sts and last 7 sts in **Seed Stitch** as set and middle 216 (238, 258, 280) sts in st st as set, ending with RS facing for next row.

Make Buttonhole
Next Row (RS) (Begin Buttonhole): Work 2 sts in **Seed Stitch** as set, BO 3 sts, work 2 sts in **Seed Stitch** as set; continue as set to end of row.

Next Row (WS) (Complete Buttonhole): Work as set to last 4 sts; work 2 sts in **Seed Stitch** as set, CO 3 sts, work 2 sts in **Seed Stitch** as set.

Hereafter, make buttonholes approx. every 4", making last (5th) buttonhole 2 rows before neck shaping.

Next Row (RS): Work first 7 sts in **Seed Stitch** as set; k9 (4, 8, 8), ([k2tog, k7 (8, 9, 9)] 22 (23, 22, 24) times), k9 (4, 8, 8); work rem 7 sts in **Seed Stitch** as set (208 (229, 250, 270) sts on needle).

Next Row (WS): Work first 7 sts in **Seed Stitch** as set; p194 (215, 236, 256); work rem 7 sts in **Seed Stitch** as set.

Leave sts on needle.

Tier 3
With another needle, CO 194 (215, 236, 256) sts and work 6 rows in **Seed Stitch**, ending with RS facing for next row.

Join Tier 2 to Tier 3

Next Row (RS): Work first 7 sts of **Tier 2** in **Seed Stitch** as set; place **Tier 3** needle on top of **Tier 2** needle and knit the tiers tog; work rem 7 sts of **Tier 2** in **Seed Stitch** as set (208 (229, 250, 270) sts on needle).

Continue on **Tier 3** as set for 55 more rows, always working first 7 sts and last 7 sts in **Seed Stitch** as set and middle 194 (215, 236, 256) sts in st st as set, ending with RS facing for next row.

Next Row (RS): Work first 7 sts in **Seed Stitch** as set; k7 (5, 10, 11), ([k2tog, k8 (8, 7, 7)] 18 (21, 24, 26) times), k7 (0, 10, 11); work rem 7 sts in **Seed Stitch** as set (190 (208, 226, 244) sts on needle).

Next Row (WS): Work first 7 sts in **Seed Stitch** as set; p176 (194, 212, 230); work rem 7 sts in **Seed Stitch** as set.

Leave sts on needle.

Tier 4

With another needle, CO 176 (194, 212, 230) sts and work 6 rows in **Seed Stitch**, ending with RS facing for next row.

Join Tier 3 to Tier 4

Next Row (RS): Work first 7 sts of **Tier 3** in **Seed Stitch** as set; place **Tier 4** needle on top of **Tier 3** needle and knit the tiers tog; work rem 7 sts of **Tier 3** in **Seed Stitch** as set (190 (208, 226, 244) sts on needle).

Continue on **Tier 4** as set for 13 more rows, always working first 7 sts and last 7 sts in **Seed Stitch** as set and middle 176 (194, 212, 230) sts in st st as set, ending with RS facing for next row.

****

Both jacket and coat versions are worked the same from this point to end of pattern.

Divide for Fronts and Back

Next Row (RS): Work 49 (54, 59, 63) sts (right front); place next 92 (100, 108, 118) sts on holder for back; place rem 49 (54, 59, 63) sts on holder for left front.

Right Front
Shape Underarm

Continuing on right front only, BO at armhole edge 5 sts once, 3 sts once, 2 sts 1 (1, 1, 2) time(s), then dec 1 st 5 (9, 10, 10) times (34 (35, 39, 41) sts rem on right front). Continue without further shaping until armhole measures 5½, ending with RS facing for next row. Work final buttonhole on next 2 rows. Work 2 more rows.

Shape Right Neck

Continuing as set, BO at neck edge 6 (7, 8, 9) sts once, 3 (3, 4, 5) sts once, 2 sts twice, then dec 1 st 3 (4, 4, 4) times. Work without further shaping on rem 18 (17, 19, 19) sts until armhole measures 8½", ending with RS facing for next row.

Shape Right Shoulder

Next Row (RS): K9 (9, 10, 10), RS W&T, purl back.
Next Row (RS): Knit to wrapped st, then knit wrap tog with st to hide it. Knit to end of row.

Place 18 (17, 19, 19) shoulder stitches onto holder.

Back

Move 92 (100, 108, 118) sts on holder for back onto needle and rejoin yarn at right armhole. Working in st st, BO 5 sts at beg of next 2 rows, 3 sts at beg of next 2 rows, 2 sts at beg of next 2 (2, 2, 4) rows, then dec 1 st at beg and end of every RS row 5 (9, 10, 10) times. Work without further shaping on rem 62 (62, 68, 74) sts until back is 4 rows shorter than right front, ending with RS facing for next row.

Shape Back Neck and Left Shoulder

Next Row (RS): K18 (17, 19, 19), BO next 26 (28, 30, 36) sts for back neck, k18 (17, 19, 19).

Continue on left shoulder only.

Next Row (WS): Purl.
Next Row (RS): K9 (9, 10, 10), RS W&T, purl back.
Next Row (RS): Knit to wrapped st, then knit wrap tog with st to hide it. Knit to end of row.

Place 18 (17, 19, 19) shoulder sts onto holder.

Shape Right Shoulder

Rejoin yarn to right shoulder at neck edge.

Next Row (WS): P9 (9, 10, 10), WS W&T, knit back.
Next Row (WS): Purl to wrapped st, then purl wrap tog with st to hide it.

Place 18 (17, 19, 19) shoulder sts onto holder.

Left Front

Move 49 (54, 59, 63) sts on holder for left front onto needle and work same as for right front, reversing shaping and omitting buttonholes.

Join Shoulders
Join shoulders using 3-needle bind-off method.

Collar
With RS facing, beg 3 sts from edge, pick up 20 (22, 24, 26) sts up right front neck edge to shoulder, 33 (35, 37, 43) sts along back neck edge, and 20 (22, 24, 26) sts down left front neck edge, ending 3 sts from edge (73 (79, 85, 95) sts on needle).

Rows 1 & 3 (WS): K1, *p1, k1; rep from *.
Rows 2 & 4 (RS): P1, *k1, p1; rep from *.

Change to **Seed Stitch**, and inc 1 st at beg and end of every 4th row until collar measures 3¼" from beg of **Seed Stitch**, ending with RS facing for next row.

BO 3 sts at beg of next 2 rows, 2 sts at beg of next 4 rows, then dec 1 st at beg and end of next 4 rows. Work 1 row. BO rem sts.

Sleeves
Lower Sleeve
CO 128 (140, 152, 164) sts. Work in **Seed Stitch** for 6 rows, ending with RS facing for next row. Work in st st until piece measures 9" from CO edge, ending with RS facing for next row.

Next Row (RS): K2tog along entire row (64 (70, 76, 82) sts rem).
Next Row (WS): Purl.

Leave sts on needle and set aside.

Upper Sleeve
With another needle, CO 64 (70, 76, 82) sts. Work in **Seed Stitch** for 6 rows, ending with RS facing for next row.

Join Upper Sleeve to Lower Sleeve
Next Row (RS): Place **Upper Sleeve** needle on top of **Lower Sleeve** needle and knit the pieces tog.

Continue in st st on **Upper Sleeve** for 13 more rows, ending with RS facing for next row.

Shape Armhole
BO 5 sts at beg of next 2 rows, 3 sts at beg of next 2 rows, 2 sts at beg of next 2 (2, 2, 4) rows, then dec 1 st at beg and end of every RS row 5 (9, 10, 10) times. Work without further shaping on rem 34 (32, 36, 38) sts until **Upper Sleeve** measures 7½" from beg of armhole shaping. BO 2 sts at beg of next 4 rows and 3 sts at beg of next 4 rows. BO rem 14 (12, 16, 18) sts.

Finishing
Sew sleeves into armholes. Sew sleeve seams. Block gently. With D/3 crochet hook, work a row of single crochet around collar edge. Sew on buttons opposite buttonholes.

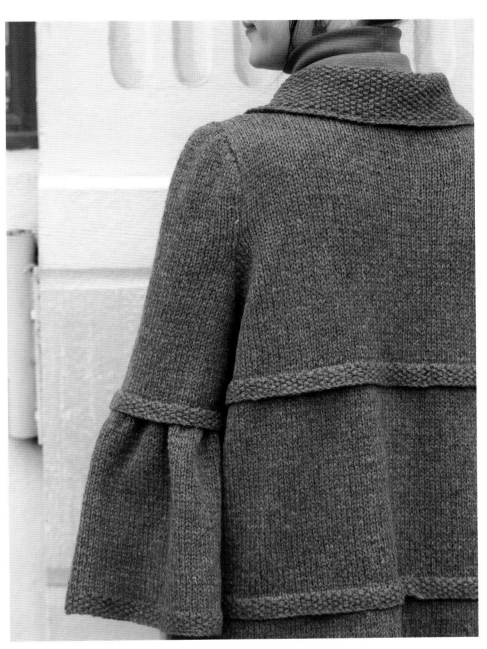

rosalie scarf

gregory courtney

Materials
Yarn: Simply Shetland Lambswool & Cashmere - 50 grams. Shown in Olive Grove (1057).
Needles: US 6 (4 mm), *or correct needles to obtain gauge*.
Accessories: Stitch holders.

Measurements
Width of Lace Portion: 4". **Width of Strap Portion:** 2". **Length:** 41".

Gauge
On US 6 in **Lace Pattern (unblocked):** 25 sts and 35 rows = 4".

Designer Note
Make the scarf in two identical halves, then join at center (back neck).

Lace Pattern (Multiple of 12 + 1)
Rows 1-4: Knit.
Rows 5, 7, 9 & 11 (RS): K1, *([k2tog] twice), ([yo, k1] 3 times), yo, ([ssk] twice), k1; rep from *.
Rows 6, 8, 10 & 12 (WS): Purl.

Rep Rows 1-12.

Scarf Half
CO 25 sts. Work the 12 rows of **Lace Pattern** 3 times.

Next 4 Rows: Knit.

Decrease for Strap
Next Row (RS): ([Sl 1 kwise wyib] twice), k1, ssk; **knit** to last 5 sts; k2tog, sl 1 pwise wyib, k1, p1.
Next Row (WS): Sl 1 kwise wyib, sl 1 pwise wyif, p1; **purl** to last 3 sts; sl 1 pwise wyif, p1 tbl, p1.

Rep these last 2 rows until 15 sts rem on needle, ending with RS facing for next row.

Work Strap
Next Row (RS): ([Sl 1 kwise wyib] twice), k1; **knit** to last 3 sts; sl 1 pwise wyib, k1, p1.
Next Row (WS): Sl 1 kwise wyib, sl 1 pwise wyif, p1; **knit** to last 3 sts; sl 1 pwise wyif, p1 tbl, p1.

Rep these last 2 rows until scarf measures 20½" from CO edge, ending with RS facing for next row. Place sts on holder. Work **Scarf Half** again.

Finishing
Weave halves together or join using using 3-needle bind-off method. Block gently, being careful not to lose the gentle undulation in the lace pattern.

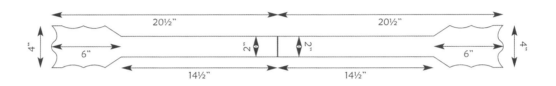

leaf poncho jacket

nicky epstein

MATERIALS
YARN: Jamieson's Shetland Heather Aran - 1,000 grams. Shown in Pippin (808).
NEEDLES: 32" circular and set of 2 double-pointed US 7 (4.5 mm), *or correct needles to obtain gauge.*
ACCESSORIES: Stitch holders.

MEASUREMENTS
CHEST: 64".
LENGTH (NOT INCLUDING SCALLOPED EDGING): 28".
SLEEVE LENGTH: 16½".

GAUGE
On US 7 in st st: 17 sts and 26 rows = 4".

NOTES ON CHARTS
Work odd-numbered (RS) rows from right to left, and even-numbered (WS) rows from left to right.

SCALLOPED EDGING
Row 1 (WS): Knit.
Row 2 (RS): Purl.
Row 3: Knit.
Row 4: K1; *yo, k21; rep from * to last 2 sts; end k2.
Row 5: P2; *([p1, k3] 5 times), p2; rep from * to last st; end p1.
Row 6: K1; *k1, yo, k1, ([p3, k1] 5 times), yo; rep from * to last 2 sts; end k2.
Row 7: P1; *p3, ([k3, p1] 5 times), p1; rep from * to last 2 sts; end p2.
Row 8: K1; *([k1, yo] twice), ([ssk, p2] 5 times), ([k1, yo] twice); rep from * to last 2 sts; end k2.
Row 9: P2; *p4, ([k2, p1] 5 times), p4; rep from * to last st, end p1.
Row 10: K1; *([k1, yo] 4 times), ([ssk, p1] 5 times), ([k1, yo] 4 times); rep from * to last 2 sts; end k2.
Row 11: P2, *p8, ([k1, p1] 5 times), p8; rep from * to last st; end p1.
Row 12: K1, *k8, ([ssk] 5 times), k8; rep from * to last 2 sts; end k2.
Row 13: P2; *p8; p4 and sl these sts to cn, wrap yarn clockwise around sts on cn 3 times, sl these sts to right-hand needle, p9; rep from * to last st; end p1.

BACK
CO 150 sts. Work Rows 1-12 of **Scalloped Edging**. At end of Row 12, CO 6 sts (156 sts on needle).

Next Row (WS): P6; work Row 13 of **Scalloped Edging**; CO 6 sts (162 sts on needle).

Next Row (RS): Work Row 1 of **Chart A**; *k9, work Row 1 of **Chart A**; k6; rep from * to last 9 sts; k3, work Row 1 of **Chart A**.

Continue as set, repeating the 4 rows of **Chart A** and working st st between, until piece measures 28", not including **Scalloped Edging**, ending with RS facing for next row.

Last Row (RS): Work 65 sts as set and place on holder for right shoulder; BO 32 sts for back neck; work rem 65 sts as set and place on holder for left shoulder.

RIGHT FRONT
CO 66 sts. Work Rows 1-12 of **Scalloped Edging**. At end of Row 12, CO 6 sts.

Next Row (WS): P6, work Row 13 of **Scalloped Edging**; CO 16 sts (88 sts on needle).

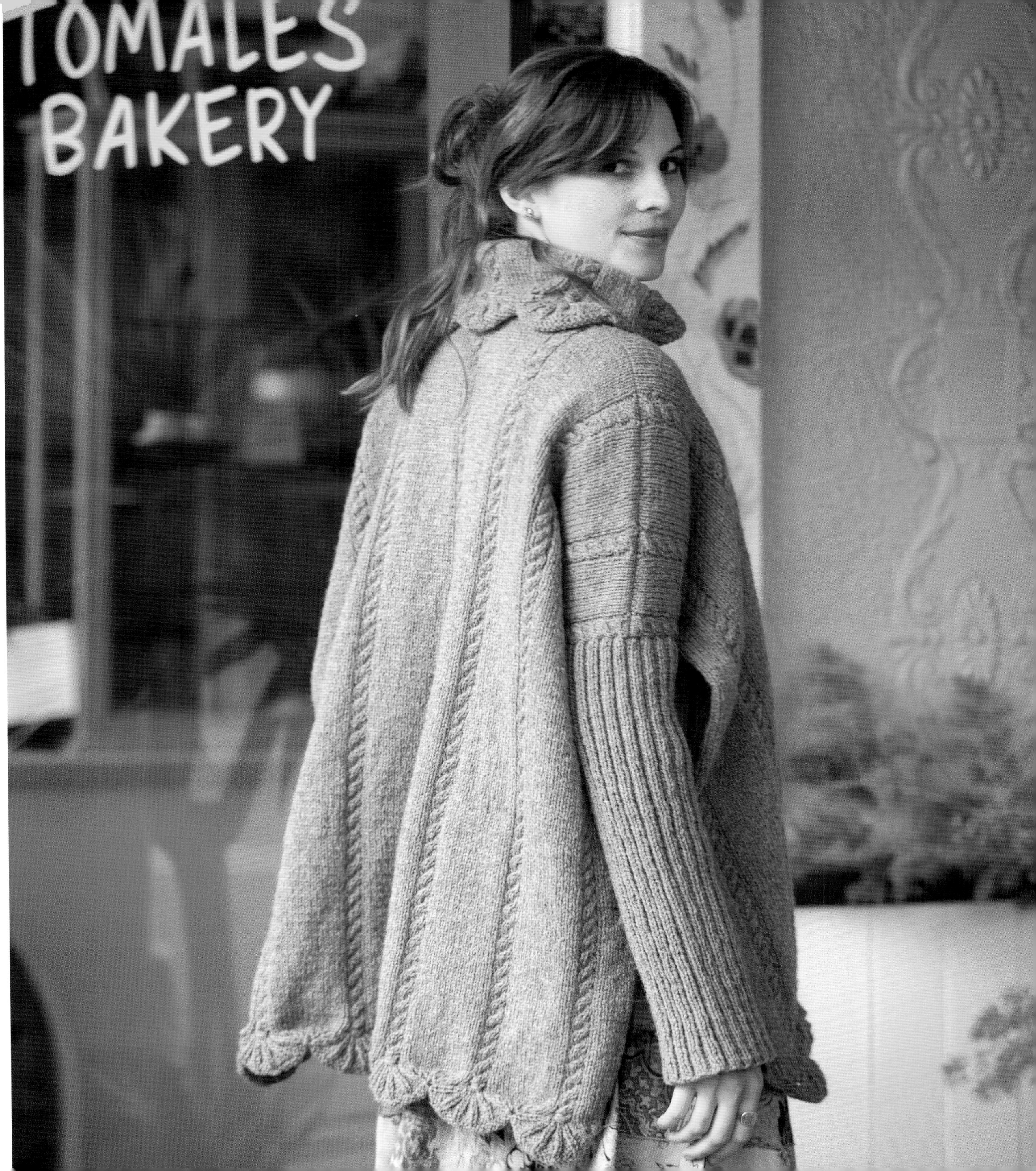
TOMALES
BAKERY

leaf poncho jacket

Next Row (RS): Work Row 1 of **Chart B**; *k9, work Row 1 of **Chart A**, k6; rep from * to last 9 sts; k3, work Row 1 of **Chart A**.

Continue as set, repeating the 4 rows of **Chart A** and **Chart B** and working st st between, until piece measures 26½", excluding **Scalloped Edging**, ending with RS facing for next row.

Shape Right Front Neck
Next Row (RS): BO 16 sts; work as set to end of row.

Continuing as set, BO 2 sts at neck edge every RS row twice, then dec 1 st at neck edge on every RS row 3 times, ending after working a RS row. Place rem 65 sts on holder for right shoulder.

Left Front
CO 66 sts. Work Rows 1-12 of **Scalloped Edging**. At end of Row 12, CO 16 sts.

Next Row (WS): P16; work Row 13 of **Scalloped Edging**, CO 6 sts (88 sts on needle).

Next row (RS): Work Row 1 of **Chart A**; *k9, work Row 1 of **Chart A**, k6; rep from * to last 19 sts; k3, Row 1 of **Chart B**.

Continue as set, repeating the 4 rows of **Chart A** and **Chart B** and working st st between, until piece is same length as right front up to neck shaping, excluding **Scalloped Edging**, ending with WS facing for next row.

Shape Left Front Neck
Next Row (WS): BO 16 sts; work as set to end of row.

Continuing as set, BO 2 sts at neck edge on every WS row twice, then dec 1 st at neck edge on every WS row 3 times. Place rem 65 sts on holder for left shoulder.

Join Shoulders
Join shoulders using 3-needle bind-off method.

Sleeves
Place markers along edges 6½" down from shoulder seams. With RS facing, pick up 80 sts along edge between markers.

Every Row: *K2, p2; rep from *.

When sleeve measures 16½", BO. Sew sleeve seams. Leave side seams unsewn.

leaf poncho jacket

Collar
CO 66 sts. Work Rows 1-11 of **Scalloped Edging**. At end of Row 11, CO 16 sts.

Next Row (RS): K16; work Row 12 of **Scalloped Edging**; CO 16 sts (98 sts on needle).

Next Row (WS): ([K1, p4] 3 times), k1; work Row 13 of **Scalloped Edging**, ([k1, p4] 3 times), k1.

Next Row: Work Row 1 of **Chart B**; *k9, work Row 1 of **Chart A**, k6; rep from * to last 19 sts; k3, work Row 1 of **Chart B**.

Continue as set, repeating the 4 rows of **Chart A** and **Chart B** and working st between, until collar measures 5½", excluding **Scalloped Edging**. BO. Sew collar to neck edge.

Leaf Frogs
Each leaf frog consists of a knitted cord with a leaf "growing" out of each end. Make 5 frogs with a 6" cord and 5 with a 3" cord.

Knitted Cord
With double-pointed needles, CO 3 sts and work as follows: *k3; without turning work, slide sts to opposite end of needle, pull yarn across back of sts; rep from * to desired length. Without turning, leave sts on needle and continue into **Leaf** as follows:

Leaf
Row 1 (RS): K1, yo, k1, yo, k1 (5 sts on needle).
Row 2 and all even-numbered (WS) rows: Purl.
Row 3: K2, yo, k1, yo, k2 (7 sts on needle).
Row 5: K3, yo, k1, yo, K3 (9 sts on needle).
Row 7: Ssk, k5, k2tog (7 sts on needle).
Row 9: Ssk, k3, k2tog (5 sts on needle).
Row 11: Ssk, k1, k2tog (3 sts on needle).
Row 13: K3tog; draw thread through rem loop and fasten off.

Work Another Leaf at Opposite End of Knitted Cord
With RS of previous leaf facing, slide needle into sts at CO end of knitted cord (3 sts on needle). Without turning, reattach yarn and work **Leaf** instructions above.

Finishing
Divide leaf frogs into two groups: those with 6" cords and those with 3" cords. Fold 6" cords in half and tie an overhand knot at folded end of doubled cord. This knot forms the "button." The frogs with 3" cords form the loops. Sew frogs in place as shown in schematic. Position first frog ½" down from collar edge, and third frog 7" up from bottom edge. Place second frog evenly between these two. Sew one frog to each side edge 7" up from bottom.

Chart A
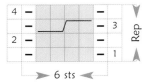

Chart B
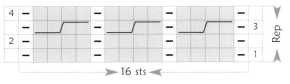

Key
- k on right side rows; p on wrong side rows.
- p on right side rows; k on wrong side rows.
- sl 2 sts to cn and hold at back; k2; k2 from cn.

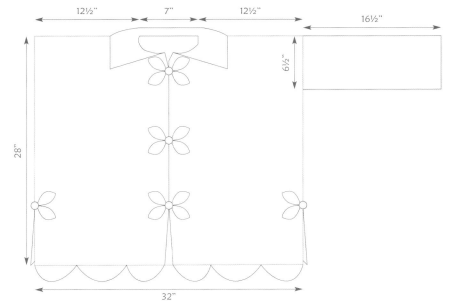

carnivale cropped top

carol lapin

MATERIALS
Yarn: Jamieson's Shetland Double Knitting - 200 grams each of Granny Smith (1140) and Chartreuse (365); 25 grams each of Black (999), Cobalt (684), Maroon (595), Mustard (425) and Spruce (805).
Needles: US 4 (3.50 mm), US 5 (3.75 mm) and US 6 (4 mm), *or correct needles to obtain gauge*. 16" circular US 6 (4 mm).

MEASUREMENTS
Chest: 44".
Length (includes color block hem band, after hemming): 18".
Sleeve Length: 14½".

GAUGE
On US 6 in st st: 22 sts and 30 rows = 4".

FRONT
Color Block Hem Band
With US 5, CO 13 sts. Working in st st throughout and always slipping the 7th st pwise on RS rows and purling it on WS rows (this will be the fold line when you hem the bottom later on), and knitting ends in as you go, ([work 6 rows each in Mustard, Black, Maroon, Cobalt and Spruce] 4 times), then work 6 rows each in Mustard, Black, Maroon and Cobalt (24 blocks in all). BO.

Main Front
With US 6 and Granny Smith, RS facing, pick up 120 sts evenly along long edge of color block hem band (5 sts in each color block).

Next Row (WS): With Granny Smith, purl.

Continuing in st st, alternate 2 rows of Chartreuse with 2 rows of Granny Smith until piece measures 9" (not including color block hem band), ending after completing a Chartreuse stripe and with RS facing for next row.

Continue into Sleeves
Next Row (RS): With Granny Smith, work to end of row; CO 28 sts, change to Mustard and CO 7 sts.

Next Row (WS): With Mustard, work 7 sts, with Granny Smith, work to end of row, CO 28 sts, change to Cobalt and CO 7 sts (190 sts on needle).

Continue Chartreuse/Granny Smith stripes and work color block sleeve bands in color sequence as for color block hem band until piece measures 12" (not including color block hem band), ending with RS facing for next row.

Shape Front Neck
Next Row (RS): Work to 3 sts before center, k2tog, k1.

Turn, and working left front of neck only, dec as above at neck edge on every RS row until 73 sts rem. Continue without further shaping until piece measures 17" (not including color block bottom band), ending after working a RS row. You'll have 10 color bands in the color block sleeve bands (last color will have 1 extra row). Place

carnivale cropped top

shoulder sts on holder. Rejoin yarn and work dec's for right front of neck as follows:

Next Row (RS): K1, ssk, work to end of row.

Continuing as set, dec at neck edge as above on every RS row until 73 sts rem. Continue same as for left front of neck.

Back
Work same as for front; except begin color block sleeve band with Cobalt on first end and Mustard on opposite end *(this will ensure you don't end up with adjoining blocks of the same color when you join the shoulders).*

Shape Back Neck
When piece is one row less than front, work last row as follows:

Next Row (RS): K73, BO 44, K73.

Join Shoulders
Join shoulders using 3-needle bind-off method.

Sleeve Puffs
With US 6 and Granny Smith, RS facing, pick up sts along edge of color block sleeve band as follows: *pick up 2 sts, m1; rep from * along edge until there are 160 sts on needle. *It really doesn't matter that you pick up exactly 160 sts; just be sure it's an even number.*

Next Row (WS): Purl.

Work in st st, alternating 2 rows of Chartreuse with 2 rows of Granny Smith until sleeve puff measures 7", ending with RS facing for next row. Change to US 4 and k2tog along row (80 sts rem—or half the number you actually picked up).

Next Row (WS): Purl.

Continuing in st st, work 2 rows each in Cobalt, Spruce, Mustard, and Maroon, then work 1 row in Black.

Next Row (WS) (Turning Ridge): Knit.

Continuing in st st, work 2 rows each in Cobalt, Spruce, Mustard, Maroon, and Black. BO.

Neckband
With 16" circular US 6 and Granny Smith, RS facing, beg at right shoulder seam, pick up 47 sts along back neck edge, 45 sts down left neck edge and 45 sts up right neck edge. Join and knit 4 rounds, BO.

Finishing
Turn color block hem band under at fold line and sew down. Turn sleeve cuff at turning ridge and sew down. Sew side and sleeve seams. Weave in ends. Block to finished measurements.

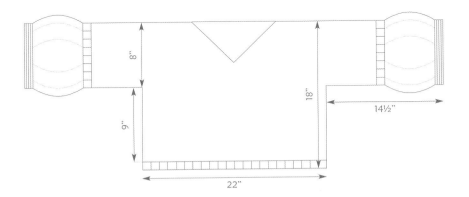

silk diamonds scarf

zara murken

MATERIALS
Yarn: Simply Shetland Silk & Lambswool - 50 grams each of Color A, Craignish (030); Color B, Melgund (011); Color C, Pitreavie (015); Color D, Kildrummy (019); and Color E, Ardvreck (034).
Needles: US 5 (3.75 mm), *or correct needles to obtain gauge*.

MEASUREMENTS
Width: 13½".
Length (including points; not including tassels): 65".

GAUGE
On US 5: one diamond measures 4½" wide x 5" long.

SCARF
Study schematic on facing page for color placement and diamond types. Instructions for the four diamond types appear below. Begin by knitting the 3 **Foundation Diamonds** at bottom of schematic *(knit each separately)*. Lay 2 of the diamonds side-by-side with RS's facing and work a **Connecting Diamond**. Lay the 3rd **Foundation Diamond** alongside the two that are now joined, and work another **Connecting Diamond**. Now continue upwards, picking up sts along tops of diamonds to work **Connecting Diamonds, Right Half Diamonds** and **Left Half Diamonds**. When all diamonds are complete, pick up 36 sts along tops of each of the last 3 diamonds (in their respective colors), and using intarsia method, work 3 garter ridges. BO. Weave in ends neatly between diamonds on WS. Block gently to finished measurements. If desired, make tassels and tie to short ends.

FOUNDATION DIAMOND
With CC, CO 36 sts, placing marker between the 18th & 19th sts and work as follows:

Row 1 (WS): With CC, knit.
Row 2 (RS): With CC, knit to 2 sts before marker, k2tog, slip marker, ssk, knit to end of row (34 sts rem).
Row 3: With CC, knit.
Row 4: With MC, knit to 2 sts before marker, k2tog, slip marker, ssk, knit to end of row (32 sts rem).
Row 5: With MC, purl.
Row 6: With MC, knit to 2 sts before marker, k2tog, slip marker, ssk, knit to end of row (30 sts rem).
Row 7: With MC, purl.
Row 8: With CC, knit to 2 sts before marker, k2tog, slip marker, ssk, knit to end of row (28 sts rem).
Row 9: With CC, knit.
Rows 10-15: Rep Rows 4-9 (22 sts rem after Row 14).
Rows 16-21: Rep Rows 4-9 (16 sts rem after Row 20).
Rows 22-27: Rep Rows 4-9 (10 sts rem after Row 26).
Rows 28-31: Rep Rows 4-7 (6 sts rem after Row 30).
Rows 32-33: Rep Rows 8-9 (4 sts rem after Row 32).
Row 34: With MC, k2tog, ssk, pass 1st st over 2nd st. Pull yarn through last rem st.

CONNECTING DIAMOND
With CC, RS facing, pick up 36 sts evenly along top edges of diamonds below. Work Rows 1-34 same as for **Foundation Diamond**.

silk diamonds scarf

 RIGHT HALF DIAMOND
With CC, RS facing, pick up 18 sts evenly along right-slanting top edge of diamond below and work as follows:

Row 1 (WS): With CC, knit.
Row 2 (RS): With CC, ssk, knit to end of row (17 sts rem).
Row 3: With CC, knit.
Row 4: With MC, ssk, knit to end of row (16 sts rem).
Row 5: With MC, purl.
Row 6: With MC, ssk, knit to end of row (15 sts rem).
Row 7: With MC, purl.
Row 8: With CC, ssk, knit to end of row (14 sts rem).
Row 9: With CC, knit.
Rows 10-15: Rep Rows 4-9 (11 sts rem after Row 14).
Rows 16-21: Rep Rows 4-9 (8 sts rem after Row 20).
Rows 22-27: Rep Rows 4-9 (5 sts rem after Row 26).
Rows 28-31: Rep Rows 4-7 (3 sts rem after Row 30).
Row 32: With CC, ssk, k1 (2 sts rem).
Row 33: With CC, knit.
Row 34: With MC, ssk. Pull yarn through last rem st.

 LEFT HALF DIAMOND
With CC, RS facing, pick up 18 sts evenly along left-slanting top edge of diamond below and work as follows:

Row 1 (WS): With CC, knit.
Row 2 (RS): With CC, knit to last 2 sts, k2tog (17 sts rem).
Row 3: With CC, knit.
Row 4: With MC, knit to last 2 sts, k2tog (16 sts rem).
Row 5: With MC, purl.
Row 6: With MC, knit to last 2 sts, k2tog (15 sts rem).
Row 7: With MC, purl.
Row 8: With CC, knit to last 2 sts, k2tog (14 sts rem).
Row 9: With CC, knit.
Rows 10-15: Rep Rows 4-9 (11 sts rem after Row 14).
Rows 16-21: Rep Rows 4-9 (8 sts rem after Row 20).
Rows 22-27: Rep Rows 4-9 (5 sts rem after Row 26).
Rows 28-31: Rep Rows 4-7 (3 sts rem after Row 30).
Row 32: With CC, k1, k2tog (2 sts rem).
Row 33: With CC, knit.
Row 34: With MC, k2tog. Pull yarn through last rem st.

silk diamonds scarf

Color Key
A—Craignish (030) (Rust/Light Green)
B—Melgund (011) (Light Green)
C—Pitreavie (015) (Medium Green)
D—Kildrummy (019) (Dark Green)
E—Ardvreck (034) (Rust)

First letter in diamond is CC (garter ridges)
Second letter is MC (st st/background)
For example, D/B = Kildrummy (CC)/Melgund (MC)

Foundation Diamond

Connecting Diamond

Right Half Diamond

Left Half Diamond

autumn rose pullover

eunny jang

MATERIALS

YARN: Jamieson's 2-Ply Shetland Spindrift - 25 (25, 25, 50, 50) grams of Shetland Black (101); 25 (25, 50, 50, 50) grams each of Pine Forest (292), Admiral Navy (727), Peat (198), Bracken (231), Pistachio (791), Yellow Ochre (230) and Scotch Broom (1160); 50 (50, 75, 75, 75) grams of Sunrise (187); 25 (25, 25, 25, 25) grams of Madder (587); 75 (75, 75, 100, 100) grams of Old Gold (429).
NEEDLES: 29" and 16" circular and set of 5 double-pointed US 2 (3 mm); 16" circular US 1 (2.5 mm); and 16" circular US 0 (2 mm), *or correct needles to obtain gauge.*
ACCESSORIES: Stitch markers. Stitch holders. Tapestry needle.

MEASUREMENTS
CHEST: 35 (37, 39, 41, 43)".
LENGTH: 22¼ (23, 23¾, 24½, 25)".
SLEEVE LENGTH (TO UNDERARM): 9¾ (10, 10¼, 10½, 10¾)".

GAUGE
On US 2 in colorwork pattern: 30 sts and 32 rows = 4".

ABOUT CHARTS
Body Chart shows half of full length of front and back; center st is not repeated. **Sleeve Chart** shows half of full length of sleeve, folded at shoulder line; center st is not repeated.

BODY
With 29" US 2 and Old Gold, CO 256 (272, 296, 304, 320) sts. Join and work ribbing in the rnd as follows, adding in and breaking off colors as needed:

Rnds 1-5: *K3 Sunrise, p1 Old Gold; rep from *.
Rnds 6-10: *K3 Madder, p1 Old Gold; rep from *.
Rnds 11-15: *K3 Sunrise, p1 Old Gold; rep from *.

Set-up Rnd: *With Old Gold only, k42 (45, 49, 50, 53), m1 (1, 0, 1, 1), k43 (45, 49, 51, 53), m1 (1, 0, 1, 1), k42 (45, 49, 50, 53); place marker, p1 ("seam st"), place marker; rep from * for back (260 (276, 296, 308, 324) sts on needle).

Work first 4 rnds of **Body Chart** for your size, working "seam sts" as set in background color on these rows and throughout garment.

Dec Rnd: *Continuing chart as set, k2tog, work to last 2 sts before first marker, ssk, slip marker, p1, slip marker; rep from * for back (4 sts dec'd).

Continuing chart as set, work **Dec Rnd** every 4th round 2 (2, 6, 7, 9) more times, then every 3rd rnd 8 (8, 4, 3, 1) times (44 sts dec'd; 216 (232, 252, 264, 280) sts rem).

Continuing chart as set, work 7 rnds without shaping.

Inc Rnd: *Continuing chart as set, m1, work to marker, m1, slip marker, p1, slip marker; rep from * for back (4 sts inc'd).

Continuing chart as set, work **Inc Rnd** every 5th rnd 6 (6, 4, 3, 1) times more, and then every 6th rnd 4 (4, 6, 7, 9) times (44 sts inc'd; 260 (276, 296, 308, 324) sts rem).

Continuing chart as set, work 5 rnds without shaping, break yarns.

ALSO AT SAME TIME, begin working front neck shaping on same rnd as 9th (10th, 10th, 10th, 10th) **Inc Rnd** by slipping center 29 (25, 31, 37, 43) sts of front to holder. Place marker on left-hand needle, CO 10 steek sts, place marker, and continue chart as set.

Body Chart

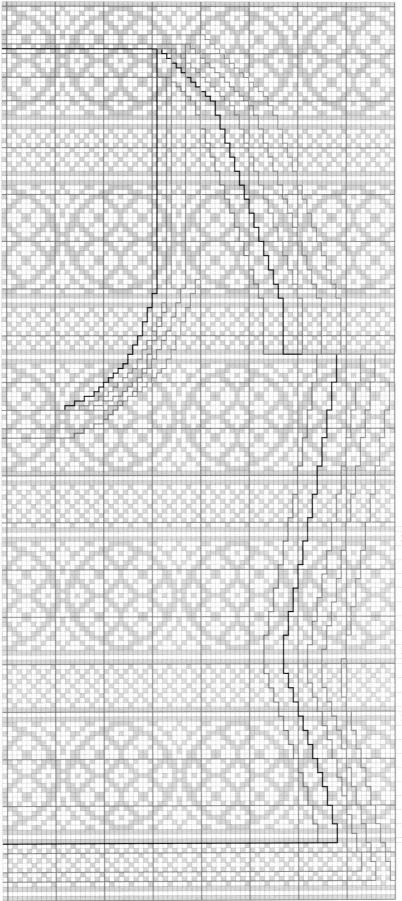

White Boxes	Black Boxes
Shetland Black	Old Gold
Pine Forest	Bracken
Admiral Navy	Pistachio
Peat	Yellow Ochre
Sunrise	Scotch Broom
Peat	Yellow Ochre
Admiral Navy	Pistachio
Pine Forest	Bracken
Shetland Black	
Sunrise	
Madder	Old Gold
Sunrise	
Shetland Black	
Pine Forest	Bracken
Admiral Navy	Pistachio
Peat	Yellow Ochre
Sunrise	Scotch Broom
Peat	Yellow Ochre
Admiral Navy	Pistachio
Pine Forest	Bracken
Shetland Black	Old Gold

Sleeve Chart

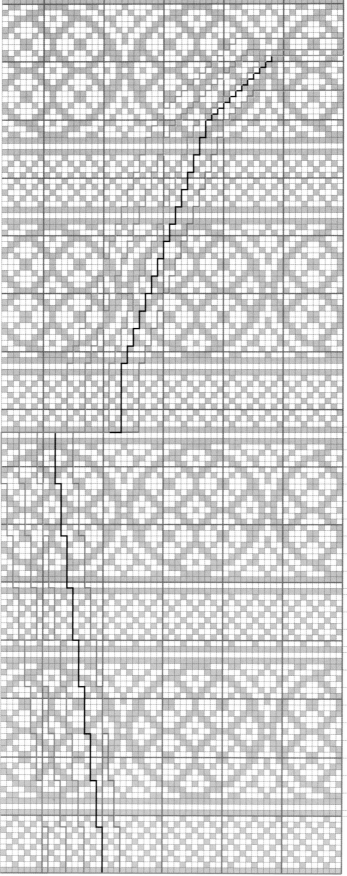

White Boxes	Black Boxes
Shetland Black	Old Gold
Pine Forest	Bracken
Admiral Navy	Pistachio
Peat	Yellow Ochre
Sunrise	Scotch Broom
Peat	Yellow Ochre
Admiral Navy	Pistachio
Pine Forest	Bracken
Shetland Black	
Sunrise	
Madder	Old Gold
Sunrise	
Shetland Black	
Pine Forest	Bracken
Admiral Navy	Pistachio
Peat	Yellow Ochre
Sunrise	Scotch Broom
Peat	Yellow Ochre
Admiral Navy	Pistachio
Pine Forest	Bracken
Shetland Black	Old Gold

autumn rose pullover

Front Neck Double Dec Rnd: Continuing side incs as set, work to within 3 sts of first front steek marker, sssk, slip marker, knit steek in stripe pattern, slip marker, k3tog. Work to end of rnd, continuing to work chart and "seam sts" as set.

Work **Front Neck Double Dec Rnd** every rnd 2 more times.

Front Neck Dec Rnd: Continuing side side incs as set, work to within 2 sts of first front steek marker, ssk, slip marker, knit steek in stripe pattern, slip marker, k2tog. Work to end of rnd, continuing to work chart and "seam sts" sts as set.

Work **Front Neck Dec Rnd** every rnd 6 (6, 5, 5, 5) more times, every other rnd 1 (3, 4, 4, 2) times, and every 3rd rnd 5 (3, 3, 3, 5) times.

NOTE: *Front neck shaping begins before and continues after body and sleeves have been joined. Discontinue work on the body of the garment after all side incs are complete and 5 rnds without shaping have been worked. Continue front neck shaping after joining body and sleeves.*

SLEEVES
With Old Gold and double-pointed US 2, CO 72 (80, 84, 92, 96) sts. Join and work ribbing in the rnd for 1½ (1¾, 2, 2¼, 2½)" as follows:

Every Rnds *K3 Sunrise, p1 Old Gold; rep from *.

Set-Up Rnd: With Old Gold only, knit to last st of rnd, inc'g 2 (0, 2, 0, 2) sts evenly along round, place marker, p1 ("seam st"). Place marker for end of rnd.

Work 1 rnd with Shetland Black only, working "seam st" as set. Work first 8 rnds of **Sleeve Chart** for your size, working "seam sts" as set in background color on these rows and throughout garment.

Inc Rnd: Continuing chart as set, m1, work to marker, m1, slip marker, p1 (2 sts inc'd).

Continuing chart as set, work **Inc Rnd** every 8th rnd 4 more times, then every 9th rnd 3 times. Work 8 rnds without shaping (90 (96, 102, 108, 114) sts). Break yarns. Slip first 10 (10, 11, 10, 10) sts and last 11 (11, 12, 11, 11) sts of rnd onto holder (21 (21, 23, 21, 21) sts on holder).

JOIN SLEEVES AND BODY
Redistribute body sts so end of rnd falls in the center of front neck steek. Reattach yarns.

Joining Rnd: Work 5 steek sts, slip marker, work across right front to 11 (11, 12, 11, 11) sts before first side marker. Slip 1 st to right-hand needle; place next 21 (21, 23, 21, 21) sts on holder. Slip unworked st on right-hand needle back to left-hand needle. Purl this body st tog with 1 st from sleeve. Work 67 (73, 79, 85, 91) sleeve sts to last st. P1 st from sleeve tog with 1 st from body. Work across back to 11 (11, 12, 11, 11) sts before second side marker. Slip 1 st to right-hand needle; place next 21 (21, 23, 21, 21) sts on holder. Slip unworked st on right-hand needle back to left-hand needle. Purl this body st tog with 1 st from sleeve. Work 67 (73, 79, 85, 91) sleeve sts to last st. P1 st from sleeve tog with 1 st from body. Work across left front to steek marker, slip marker, k5 steek sts.

Continuing to work front neck dec's as set, work across right front to first purl st ("seam st"). Place marker, p1, place marker. Work across sleeve to purl st ("seam st"). Place marker, p1, place marker. Work across back to third purl st ("seam st"). Place marker, p1, place marker. Work across sleeve to fourth purl st ("seam st"). Place marker, p1, place marker. Work across left front to end of rnd. Work 11 rnds without shaping, continuing chart as and working "seam sts" at front and back of each shoulder.

Raglan Dec Rnd: Continuing neck shaping as set, *work to 2 sts before marker, ssk, slip marker, p1, slip marker, k2tog; rep from * 3 more times. Work across left front to end of rnd.

Continuing front neck shaping as set, work **Raglan Dec Rnd** every 3rd rnd 14 (14, 13, 13, 13) more times, every 2nd rnd 0 (0, 1, 1, 2) times, and every rnd 4 (10, 11, 12, 12) times. Change to shorter circular needles and then double-pointed needles when necessary. BO all sts.

NECKBAND
With Shetland Black, graft underarms tog. Cut front neck steek carefully along center line. With Old Gold and 16" circular US 2, RS facing, beg at either shoulder seam, pick up 244 (248, 264, 284, 292) sts evenly around neck edge. Join and work in the rnd as follows, **AND AT SAME TIME**, change to 16" circular US 1 after 3 rnds, and 16" circular US 0 after 3 more rnds:

Next 10 Rnds: *K3 Sunrise, p1 Old Gold; rep from *.

With Old Gold, BO loosely.

FINISHING
Weave in ends. Wash and block carefully. When completely dry, trim steeks and tack down.

glenice wrap

victoria prewitt

MATERIALS
YARN: Simply Shetland Lambswool & Cashmere - 650 grams. Shown in Flannel (030).
NEEDLES: 32" circular US 6 (4 mm), *or correct needles to obtain gauge.*
ACCESSORIES: 2 stitch holders. Stitch markers.

MEASUREMENTS (UNBLOCKED)
WIDTH: 17¼".   **LENGTH:** 77½".

GAUGE
On US 6 in k2 p2 ribbing (slightly stretched): 27 sts and 28 rows = 4".
Ribbing pulls inward, affecting st gauge. Unblocked, st gauge on all 138 sts is 32 sts = 4".

DESIGNER NOTE
Although RS and WS rows are given, the shawl looks basically the same on both sides.

SHAWL
CO 138 stitches.

Row 1 (RS): ([K1, p1] 10 times); place marker; ([k2, p2] 24 times), k2; place marker; ([k1, p1] 10 times).
Row 2 (WS): [[P1, k1] 10 times); slip marker; ([p2, k2] 24 times), p2; slip marker; ([p1, k1] 10 times).

Rep these 2 rows until piece measures 21", ending with RS facing for next row.

MAKE ARMHOLE
Next Row (RS): Work 53 sts in pattern as set. Place rem 85 sts on holder.

Continue in pattern as set on 53 sts for 8", ending with RS facing for next row. Place these 53 sts on another holder.

Move 85 sts from other holder onto needle and continue in pattern as set for 8", ending with RS facing for next row. Break yarn. Place all 138 sts on needle, reattach yarn, and continue in pattern as set for 19½". Work 2nd armhole in same manner as first, then continue in pattern as set for 21". BO.

REINFORCE ARMHOLES
With RS facing, pick up approx. 112 sts evenly around armhole opening, then BO. Rep for opposite armhole.

Note: You may have to pick up more or fewer sts, depending on your individual tension. The point is to reinforce the armhole so it doesn't stretch out of shape without making it too tight.

FINISHING
Weave in ends. Block to finished measurements. If desired, increase width by opening up rib portion during blocking.

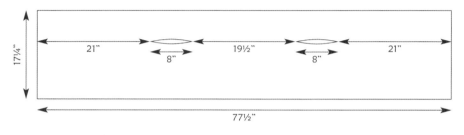

 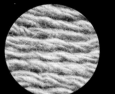  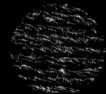  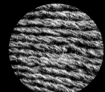  

Snow/200 Cashew/051 Flannel/030 Charcoal/190 Blueprint/273 Vista/168 Mallard/555 Kingfisher/453

Shetland Lambswool & Cashmere

87.5% Supersoft Shetland Lambswool/12.5% Cashmere
20-22 sts = 4" - US 5-6
50 Grams/Approx. 136 Yards
Spun in Scotland

 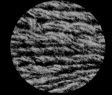 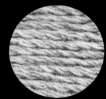 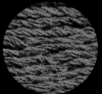   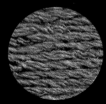 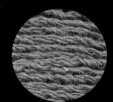

Loden/085 Olive Grove/1057 Cummin/262 Sienna/156 Rembrandt/994 Red Hot/1294 Velvet/384 Petunia/344

simply shetland—
The Colours of Scotland

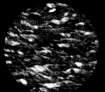 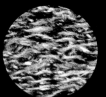 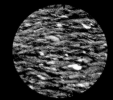 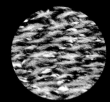 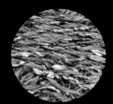 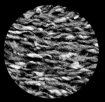  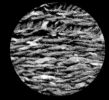

Carsluith/013 Dalcross/009 Lochmaben/014 Ethie/010 Glenbuchat/033 Moniak/029 Kildrummy/019 Pitreavie/015

Silk Noil & Shetland Lambswool

59% Silk Noil/41% Supersoft Shetland Lambswool
Single: 26-28 sts = 4" - US 2-3
Double: 18-20 sts = 4" - US 7-8
50 Grams/Approx. 246 Yards
Spun in Scotland

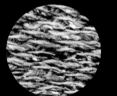 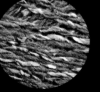 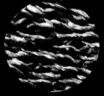  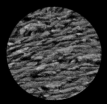 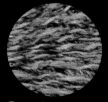 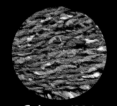

Melgund/011 Craignish/030 Slains/020 Notland/016 Ardvreck/034 Venlaw/035 Drumlanrig/031 Culzean/036

Abbreviations

alt = alternate
beg = beginning
BO = bind off
CC = contrast color
cn = cable needle
CO = cast on
dec = decrease(ing)
foll = follow(ing)
GS = garter stitch
inc = increase(ing)
k = knit
kfb = knit into front and back of st (inc)
k1b = knit through back loop
k2tog = knit 2 sts together
k2togtbl = knit 2 sts together through back loop
m1 = make 1 st (inc) - lift running thread between st just worked and next st and knit into back of loop
m1p = make 1 st (inc) purl - lift running thread between st just worked and next st and purl into front of loop
kwise = knitwise (as if to knit)
MC = main color
p = purl
p2tog = purl 2 sts together
patt = pattern
psp = p1 and slip to left-hand needle; pass next st over it; return to right-hand needle
psso = pass slipped st over st just knitted
p3sso = pass 3 slipped sts over st just knitted
pwise - purlwise (as if to purl)
rem = remaining
rep = repeat
rnd = round
RS = right side
sl = slip
ss(s)k = sl 2 (3) sts (one at a time) kwise; with left-hand needle, knit these 2 (3) sts tog through front of sts
ssp = sl 2 sts kwise (one at a time); return both sts to left-hand needle; k2togtbl
st(s) = stitch(es)
st st = stockinette stitch
tbl = through back loop
tog = together
wyib = with yarn in back
wyif = with yarn in front
WS = wrong side
yo = (inc) yarn over needle

Skill Levels

Beginner    *Intermediate*    *Expert*

Editorial Director David Codling

Editor & Graphic Design Gregory Courtney

Photography Kathryn Martin

Illustrations Molly Eckler

Historical Articles Gregory Courtney

Garments Modeled by Katie Dunn, Shawn Thornton, Jack Thornton, Michelle Rich & Cynthia Alverson

Makeup & Hair Stylist Kira Lee

Clothing Stylist Betsy Westman

A Very Special Thanks Karen Rumpza

Location Thanks The City of Tomales, California; Tomales Bakery; Tomales Regional History Center

Color Reproduction & Printing Regent Publishing Services

Jamieson's Shetland Yarns & Simply Shetland Yarns Distributed By Simply Shetland www.simplyshetland.net

Published & Distributed By Unicorn Books and Crafts, Inc. www.unicornbooks.com

Text, charts, schematics, knitwear designs, illustrations and photographs Copyright© - Unicorn Books and Crafts, Inc., 2007. All rights reserved. The knitwear designs contained in this book are not to be knitted for resale or any other commercial purpose. No part of this book may be reproduced in any way whatsoever without the written permission of the publisher.

Printed in China

ISBN 1-893063-21-6
1 2 3 4 5 6 7 8 9 10